Adsorption - Fundamental Mechanisms and Applications

Edited by Tatyana I. Shabatina and Yana Andreevna Gromova

Published in London, United Kingdom

Adsorption - Fundamental Mechanisms and Applications
http://dx.doi.org/10.5772/intechopen.1005925
Edited by Tatyana I. Shabatina and Yana Andreevna Gromova

Contributors
Adriana Vázquez-Guerrero, Alan Steven, Chenhui Zhang, Didik Prasetyoko, Fenrong Liu, Hellna Tehubijuluw, Jianbin Luo, Liudmila Yolshina, Lu Zhang, Marwah Ahmed Alsharif, Orlando Hernández-Cristobal, Rafael Huirache-Acuña, Raúl Cortés-Martínez, Reva Edra Nugraha, Riki Subagyo, Ruth Alfaro-Cuevas-Villanueva, Selene Anaid Valencia-Leal, Tatyana I. Shabatina, Tianyi Han, Varvara Dorogova, Wangjing Ma, Yana Andreevna Gromova, Yuly Kusumawati, Zhe Hang

First published in London, United Kingdom, 2025 by IntechOpen
IntechOpen is the global imprint of INTECHOPEN LIMITED, registered in England and Wales, registration number: 11086078, 167-169 Great Portland Street, London, W1W 5PF, United Kingdom

For EU product safety concerns: IN TECH d.o.o., Prolaz Marije Krucifikse Kozulić 3, 51000 Rijeka, Croatia, info@intechopen.com or visit our website at intechopen.com.

British Library Cataloguing-in-Publication Data
A catalogue record for this book is available from the British Library

Adsorption - Fundamental Mechanisms and Applications
Edited by Tatyana I. Shabatina and Yana Andreevna Gromova
p. cm.
Print ISBN 978-1-83634-491-9
Online ISBN 978-1-83634-490-2
eBook (PDF) ISBN 978-1-83634-492-6

Meet the editors

Tatyana I. Shabatina graduated with honors in 1978 from the Department of Chemistry at Moscow State University. She received her Ph.D. in Physical Chemistry from the same department in 1984 and was awarded the degree of Doctor of Chemical Sciences, specializing in Physical Chemistry, in 2013. Her international research experience includes training at the Max Planck Institute in Mülheim, Germany (1994), a research visit to Professor D. Stufkens' laboratory at the University of Amsterdam, Netherlands (1996), and another to Professor K. Klabunde's laboratory at Kansas State University, USA (2000). In 2009, she was a visiting professor at the University of York, UK. Since 2014, she has served as a Professor of Physical Chemistry and Head of the Laboratory on Low Temperature at the Chemistry Department of M. V. Lomonosov Moscow State University. Her research interests span a wide range of topics within the field of physical chemistry, with a particular focus on nanochemistry, cryochemistry of nanosized drugs, hybrid metal-organic nanosystems, chiral plasmonics, nanostructured films, low-temperature Fourier Transform Infrared (FTIR) spectroscopy, and Electron Spin Resonance (ESR) spectroscopy, as well as adsorption. These areas form the foundation of her extensive scientific contributions. She is the author of six scientific monographs and five book chapters and has served as editor or co-editor of three books. Her publication record includes 15 scientific reviews and 160 peer-reviewed scientific papers, as well as six patents registered in the Russian Federation. In addition to her research, she is actively engaged in teaching and academic supervision. She lectures on Cryochemistry of Nanoclusters and Nanostructures for graduate students at the Department of Chemistry, Moscow State University, and has supervised eight Ph.D. students and 20 graduate students throughout her academic career. Her work has been recognized with several awards, including the M. V. Lomonosov Moscow State University Prize for Young Scientists in 1984 and diplomas as a winner of the university's Innovation Projects Exhibitions in 2004, 2012, and 2016. She is also an active member of several professional organizations, including the International Scientific Committee on Low-Temperature Chemistry and Physics, the Materials Research Society (MRS), the International Liquid Crystal Society (ILCS), and the Advisory Scientific Committee of the International Conference on Low-Temperature Chemistry and Physics.

Dr. Yana Andreevna Gromova graduated from the Faculty of Chemistry at Lomonosov Moscow State University in 2014 with a Fundamental and Applied Chemistry degree. From 2014 to 2018, she pursued full-time postgraduate studies at the same faculty, specializing in Physical Chemistry. In 2018, Dr. Gromova completed her postgraduate studies with a red diploma. On November 1, 2019, she defended her dissertation and was awarded the degree of Candidate of Chemical Sciences. Currently, Dr. Gromova works as a researcher in the Laboratory of Low Temperature Chemistry within the Department of Chemical Kinetics at the Faculty of Chemistry. Her research focuses on the preparation and study of supramolecular aggregates containing Ag nanoparticles and organic mesogenic ligands such as cholesterol and thiocholesterol, including the investigation of their physicochemical properties and potential biomedical applications, particularly as drug carriers. She is also personally interested in aggregates based on Au nanoparticles and their biomedical applications, particularly in cancer therapy, with theoretical developments underway in this area. Over the years, her research has been presented at numerous international, European, and Russian conferences. In addition to her scientific work, Dr. Gromova is involved in teaching. She is a co-author of educational and methodological manuals and books and supervises coursework for both junior and senior students.

Contents

Preface

The book "Adsorption – Fundamental Mechanisms and Applications" is composed as a volume of self-contained research articles that describe different aspects of the nature, fundamental features, and mechanisms of adsorption processes occurring in natural and technological systems under various conditions, as well as some applications of such processes in science, various technical, and ecological systems. Numerous examples have demonstrated that adsorption processes have evolved into an indispensable tool for addressing various modern-day challenges, including environmental, energy, industrial, catalytic, biomedical, and other applications. The importance of adsorption lies in its diverse applications that address critical challenges in environmental protection, industrial efficiency, energy sustainability, and healthcare. Its ability to selectively remove or store substances provides a foundation for innovative technologies and systems that will improve the quality of life and reduce environmental impact.

All chapters were written by scientific experimental groups actively working in the fields of physical chemistry and the synthesis of new types of adsorbents with unique effectiveness and selectivity properties, including applications in water purification, isolation of chiral compounds, antibacterial adsorbents in biology and medicine, the development of new systems for targeted drug delivery, and various other applications.

Chapter 1, "Understanding Adsorption: Theories, Techniques, and Applications", written by Marwah Ahmed Alsharif, considers the modern meaning of "adsorption" as a complex physicochemical phenomenon and a multifaceted process that occurs in several stages depending on various factors. The chapter describes the historical aspects of the development of fundamental principles, key regulations of adsorption theories, and the modern significance of adsorption. The author discusses the fundamental mechanisms of adsorption, identifies key factors influencing the overall process and the kinetics of its various stages, and explores existing and potential applications. These include gas separation and purification, hydrogen storage for new energy sources, post-combustion carbon capture, purification of water and waste liquids, drug delivery systems, detoxification through adsorption of toxins and poisons in the gastrointestinal tract, treatment of drug overdoses, and other biomedical uses.

Chapter 2, "Selective Adsorption of Chiral Nanomaterials: Mechanisms and Biomedical Applications", by Yana Andreevna Gromova and Tatyana I. Shabatina, presents recent findings on the supramolecular synthesis of chiral nanomaterials and their biomedical applications. This chapter highlights the most promising examples of efficient adsorbents for chiral substance separation, discusses various adsorption mechanisms, and explores current and future biomedical applications. The approach involves nanomaterials capable of selectively adsorbing specific chiral isomers, offering new possibilities to enhance therapeutic medications and develop novel diagnostic methods. Research in this area is primarily focused on developing materials that enable the selective adsorption of chiral substances, which can also support targeted drug delivery by minimizing toxic effects and enhancing treatment efficacy. Thus, selective adsorption of chiral molecules represents a promising direction for the future of biomedical technology and medicinal chemistry.

The book places particular emphasis on the synthesis methods of new types of adsorbents with surfaces tailored for specific tasks. One of the most intriguing topics is the specific surface area, which determines the number of active surface centers that interact with gaseous or liquid substances and target compounds, thus influencing adsorption efficiency. Functionalization of adsorbent surfaces, whether hydrophobic or hydrophilic, makes them uniquely suited for various specialized adsorbates.

Chapter 3, "Surface Functionalization of a Biosorbent with Surfactant and Iron Nanoparticles for Fluoride Adsorption: Mechanisms and Thermodynamic Parameters", by Selene Anaid Valencia-Leal, Adriana Vázquez-Guerrero, Ruth Alfaro-Cuevas-Villanueva, Rafael Huirache-Acuña, Orlando Hernández-Cristobal and Raúl Cortés-Martínez, explores how surface functionalization using different surfactants and iron nanoparticles enables effective fluoride ion removal.

Chapter 4, "Ion Adsorption and Hydration Superlubricity", by Tianyi Han, Chenhui Zhang, and Jianbin Luo, presents experimental results on ion adsorption that lead to superlubricity, a state in which friction is nearly eliminated.

Chapter 5, "The Wettability of PMMA Surface by Aqueous Solutions of Anionic Gemini Surfactants with Different Hydrophobic Carbon Chain Lengths", by Zhe Hang, Lu Zhang, Fenrong Liu, and Wangjing Ma, investigates how the wettability of poly(methyl methacrylate) (PMMA) surfaces changes when adsorbing anionic Gemini surfactants, particularly focusing on the effect of varying hydrophobic chain lengths.

Chapter 6, "Study of Reducing and Adsorption Properties of Carbon Nanomaterials: Graphene and Hierarchically Structured Carbon Films – In Relation to Manganese Ions in an Aqueous Medium", by Liudmila Yolshina and Varvara Dorogova, presents findings on the reducing and adsorption behavior of graphene and specially synthesized carbon films when exposed to manganese ions in water.

Chapter 7, "Heterogeneous Hydrogenation in Pharmaceutical Development", by Alan Steven, showcases recent advancements in the use of heterogeneous catalytic hydrogenation for pharmaceutical process development.

Chapter 8, "Utilization of Red Mud Waste as a Dye Adsorbent", by Hellna Tehubijuluw, details the synthesis of a hierarchical zeolite adsorbent (ZSM-5) using bauxite waste (Red Mud) from Britan Island, Indonesia, in the Riau Archipelago. The two-stage synthesis yields a high-mesoporous material with a specific surface area of 734 m^2/g, which is used to remove methylene blue from water. The adsorption kinetics were shown to follow a pseudo-first-order model.

The editors hope that this book will be both interesting and useful to scientists and researchers working in the fields of physical and chemical adsorption and chromatography. It is also intended to introduce graduate and PhD students to the fascinating world of physicochemical processes involved in adsorption and its diverse, promising applications across many fields.

Tatyana I. Shabatina and Yana Andreevna Gromova
Department of Chemistry,
Lomonosov Moscow State University,
Moscow, Russian Federation

Chapter 1

Understanding Adsorption: Theories, Techniques, and Applications

Marwah Ahmed Alsharif

Abstract

The chapter topic "Understanding Adsorption: Theories, Techniques, and Applications" will provide a comprehensive exploration of the fundamental theories, experimental techniques, and practical applications of adsorption processes. It will begin by discussing the foundational theories that explain adsorption behaviour, including physisorption and chemisorption, and their underlying molecular interactions. Key adsorption models, such as Langmuir, Freundlich and BET isotherms, will be reviewed to offer insights into adsorption dynamics, capacity and surface interactions. The chapter will also cover various experimental techniques used to characterise adsorption, such as gravimetric methods, gas adsorption analysis and calorimetry, which help determine the properties of adsorbents and adsorbates. Furthermore, the chapter will examine real-world applications of adsorption in diverse fields, including environmental engineering for water and air purification, gas separation, catalysis and energy storage systems. The potential of advanced adsorbent materials, such as porous materials and nanostructures, will be discussed in light of emerging challenges and technological advancements. This chapter aims to equip readers with both theoretical knowledge and practical insights into the wide-ranging applications of adsorption processes.

Keywords: adsorption mechanisms, nanomaterials for adsorption, adsorption thermodynamics, adsorbate, adsorbent

1. Introduction

The phenomenon of adsorption is defined as the accumulation of molecules, atoms or ions from a fluid, which could be either a gas or liquid, onto the surface of a solid material [1]. The concept is different from absorption in that the latter is based on the bulk penetration of material into the surface. While this process is different, it has different nonindustrial and environmental applications, which are based on its ability to interact with specific molecules selectively. This means it allows purification, separation and catalytic transformation, and these techniques are used in applications such as water purification, gas separation, energy storage and many others. An example of this is the use of activated carbons to remove organic pollutants from water and also from volatile organic compounds found in the air [2, 3]. Within applications related to environmental sustainability, techniques for removing heavy metals from wastewater are commonly used, but there are numerous applications that can be adopted using this approach [4, 5]. The broad range of applications that the approach provides is what makes it ideal for addressing many of the challenges that are present.

1.1 Historical background and evolution of adsorption science

The concept came into the limelight in the eighteenth century when Carl Wilhelm Scheele observed that charcoal was able to absorb gases, which was then later further explored by Michael Faraday with a focus on investigating the concept of absorption related to solids and gases [6, 7]. Later in the twentieth century, with increasing interest, Irving Langmuir developed a quantitative theory linked to adsorption, which focused on monolayer adsorption on homogeneous surfaces, and this theory is considered the groundwork for modern surface chemistry and is still popular [8]. Building on this theory, the Brunauer-Emmet-Teller (BET) theory was introduced, which shifted from homogenous layers and focused on multilayer adsorption, which provided increased insight into the phenomenon of porous surfaces [9]. The theory has been used to advance technology and determine adsorbents' surface area and porosity. At the end of the twentieth century and the beginning of the twenty-first century, many advancements were observed, and in the twentieth century, the discovery of metal-organic frameworks (MOFs) could be considered a revolutionary invention. MOFs have been a focus due to their higher surface area and other properties the material possesses, like tuneable porosity and selectivity, which all make it ideal for gas adsorption applications [10, 11]. In the twenty-first century, further advances were observed, and in this, the use of nanotechnology and computational modelling has allowed for the use of materials like graphene and covalent-organic frameworks (COFs), which have shown improved capabilities in different applications like drug delivery and environmental protection [12, 13].

1.2 Scope and objectives

The focus of this particular chapter is to shed light on the concept and science behind the phenomenon and explore the existing best practices. The objectives include:

- Defining the concept and the principles behind it, including molecular interactions and thermodynamics.
- Reviewing classical adsorption isotherms like BET to understand surface interaction and adsorption capacity.
- Highlighting experimental methods like gravimetric analysis and gas adsorption isotherms to measure adsorbent properties.
- Discuss traditional and advanced adsorbents, including activated carbons, COFs and MOFs.

DOI: http://dx.doi.org/10.5772/intechopen.1008865

- Examining the role of the phenomenon in solving real-world problems.
- Addressing challenges in scaling adsorption technologies and exploring innovation in the field for sustainable adsorbent development.

2. The theories of adsorption

Understanding the mechanisms, isotherms, thermodynamics and kinetics of adsorption is critical in optimising the adsorption processes.

2.1 Fundamental mechanisms

Adsorption mechanisms are mostly categorised into physisorption and chemisorption. Physisorption has weak van der Waals forces, which would lead to low adsorption enthalpies, which are between 5 and 40 kJ/mol. The process is generally reversible and happens at very low temperatures, and it could also occur between multiple layers in the adsorbent material [14]. On the other hand, the process of chemisorption is based on the formation of strong chemical bonds between the adsorbate and the surface, and these are known to have higher enthalpies ranging between 40 and 800 kJ/mol. Chemisorption is also irreversible, and these are favoured at higher temperatures, which also require higher activation energy. The difference between the two types is considered critical for catalysis, where strength and reversibility of adsorption would impact the catalytic efficiency [14]. It is also important to note that the size, radius, polarity, shape, molecular weight and other characteristics of the component in the adsorption impact the interactions between the absorbent sources and the adsorbed particles [14]. In the case that there is more than one component within the solution, it would lead to the selective adsorption of one of these components in the solution without using the other. Increasing the molecular weight of the adsorbent is found to help in the process on the surface as a result of the increase in the possibility of its association with the surface in more than one site, and the existence of different aromatic rings in the composition is also said to impact the distribution of the surface when it comes to their nature of regulatory or homogeneity [15, 16]. **Figure 1** provides an overview of the adsorption approaches and shows the bonding formation in the techniques.

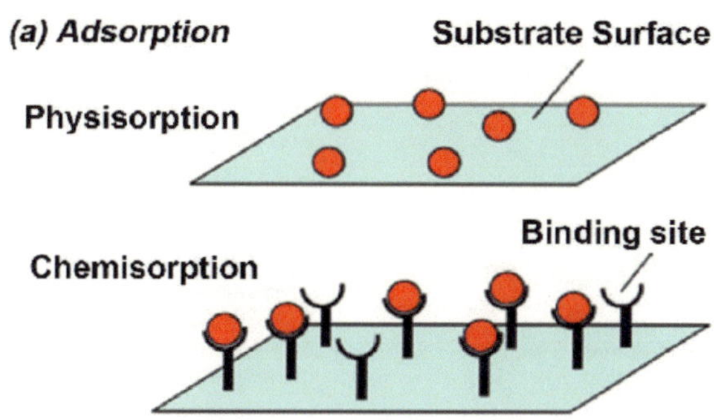

Figure 1.
Adsorption approaches [17].

There are several factors that impact the adsorption process, one of which is the nature of the gases. Elements that are easier to liquefy tend to be more adsorbable, and adsorption increases with the critical temperature of the gas. The adsorption process in studies has been shown to increase with the increase in air pressure, and the trend shown is that there is a rapid increase at first, after which it would gradually slow down as the surface would then be covered by the gas particles [18]. Another factor that has an impact is the nature of the surface area, and it is mentioned that when the surface area is large, the adoption efficiency is also large. Increasing the temperature is found to increase surface activity, and for coal, it is found to increase the surface activity when the temperature is raised from 350 to 1000 degrees celsius [14].

The adsorption mechanism involves interaction between the adsorbate molecules and the surface of the adsorbent, which is ideally represented by the Langmuir and Freundlich models [19]. According to the Langmuir model, in monolayer adsorption, within which there are active sites on the adsorbent, the adsorbate molecule would bind to one of the adsorbate molecules, which emphasises specific

Adsorbate	Solvent	Adsorption capacity (mmol/g)	Key influencing factors	Gibbs free energy change (ΔG) (kJ/mol)
Phenol	Water	3.11	Small size and lack of bulky substituents enable better micropore access and stronger interaction	−2/27
	HCl	1.61	A positive charge on adsorbent reduces interaction	−2.44
	NaOH	1.03	Repulsion Forces due to negatively charged adsorbate and adsorbent	−6.53
Salicylic Acid	HCl	1.49	Ionisation mitigates repulsion; hydrogen bonding improves adsorption	−3.39
	Water	1.44	Carboxylic acid substituent reduces π-stacking interactions and makes the process heterogenous	No data
	NaOH	0.57	Repulsion Forces dominate due to negative charges	−2.49
Methylparaben	Water	1.58	Substituents hinder power access but allow moderate adsorption based on π-stacking interactions	−5.26
	HCl	1.39	Ester groups cause displacement of water from activated carbon	−7.32
	NaOH	1.41	Stronger interaction with basic functional groups on adsorbent	−8.12

Table 1.
Adsorption capacities and mechanisms for adsorbates on activated carbon [19].

interactions like π-stacking and hydrogen bonding. Phenol is said to show the highest adsorption capacity at 3.11 mmol/g in water, largely due to the smaller size and lack of any bulky substituents, which enables better accessibility for micropores and stronger interaction with activated carbon [19]. Methylparaben, which has an adsorption capacity of 1.58 mmol/g, is said to show reduced capacities, which are hindered by the lack of pore access due to substituents and altering the chemical reactivity [19]. pH levels play a critical role as they modify the surface charge and ionisation states and, at a basic level, are said to reduce adsorption as a result of the electrostatic repulsion that is absorbed between the negatively charged adsorbent and the adsorbate species. Thermodynamically, the process was spontaneous across conditions that had Gibbs free energy changes that range from −2.27 to −8.12 kJ/mol, and the immersion enthalpy experiments show that there is a change in the energy once the interaction leads to displacement of the solvent molecules [19]. **Table 1** summarises the information and helps provide clarity on how different adsorbates react to activated carbon.

2.2 Adsorption isotherms

Adsorption isotherms describe how the amount of adsorbate on the adsorbent would vary with pressure or concentration at a constant temperature. There are many adsorption isotherm models, and in this section, three of them are discussed: Langmuir isotherm, Freundlich isotherm and BET theory.

2.2.1 Langmuir isotherm

This model is based on the assumption that the distribution of the reactive groups across the surface of the particles is homogenous, as there are no lateral interactions between the particles [20, 21]. The equation used with the model has been developed based on the theory of gases and is mostly used to describe the gas adsorption on solids. The isotherm is applicable in scenarios where solute adsorption from a solution occurs as a monolayer, characterised by a finite number of identical sites on the adsorbent and a uniform adsorption mechanism, with no observed migration of the adsorbate across the surface plane [22]. The nonlinear equation for the model is:

$$Q_e = \frac{Q_0 K_L C_e \theta}{(1 + K_L C_e)} \tag{1}$$

There is some difference when it comes to the linearised model, and it is represented as:

$$\frac{C_e}{Q_e} = \frac{1}{K_L Q_0} + \frac{C_e}{K_L} \tag{2}$$

In this, the Q_0 represents the maximum amount that is absorbed, while K_L is the Langmuir constant. The monolayer assumption is said to require identical adsorption sites, and from this, only one molecule would be adsorbed at each site. There is no more adsorption at a site when a molecule already occupies it. The model is said to convert the Henry isotherm at very low contents, and the L-shape of the curve,

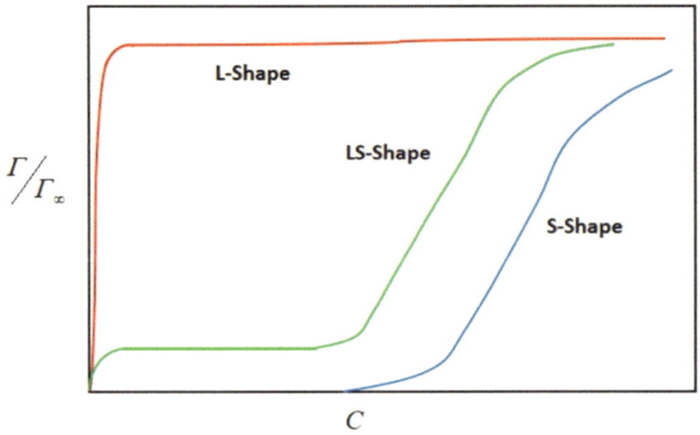

Figure 2.
L-, S- and LS- shape adsorption isotherms [23].

as shown in **Figure 2**, shows the Langmuir isotherm model having a single plateau. Another important parameter that is important for the model is the separation factor or equilibrium parameter, which is denoted as R_L and is used for checking if the adsorption is favourable or unfavourable and it is mathematically represented as:

$$R_L = \frac{1}{1 + K_L C_O} \tag{3}$$

In eq. 3, the K_l and C_0 are the Langmuir constant and the highest initial concentration of the surfactant. In general, a value of this variable, that is R_L is less than one would indicate the adsorption is favourable, and if it is equal to one, it means the adsorption is linear, and if it is greater than one, then it is unfavourable adsorption [23]. This model assumes that the adsorption occurs just at one layer or the surface, which is inaccurate when multilayer adsorption occurs. The theory also assumes that the sites of adsorption on the surface have the same energy and are equally accessible, but it fails to understand the heterogeneities that are present with real-world adsorbents. It also neglects the lateral interaction between the adsorbed molecule and the limited applicability in low adsorbate concentrating pressure [22, 23].

2.2.2 Freundlich isotherm

Unlike Langmuir, this empirical model is perfectly capable of being used in multilayer adsorption within heterogeneous sites [23]. The assumption is that the adsorption heat distribution, along with affinities towards the heterogenous surface, is found to be non-uniform, and the model can be represented using the equation:

$$q_e = bC_e^{1/n} \tag{4}$$

In the above equation, b is the adsorption capacity, while 1/n is the intensity or surface heterogeneity. When the intensity is between zero and one, it is considered favourable; if it is greater than 1, it is considered unfavourable; and lastly, if it is one, it is irreversible [23]. The linearised form of the equation is easy, but on the other hand, the linearisation process would create more errors, which would create erroneous predictions. Thus, using nonlinear regression to solve the nonlinear model is recommended for calculating the model parameters. The isotherm is capable of describing the multilayer adsorption and also assumes the exponential decay when it comes to the energy distribution of adsorbed sites, but this is not valid for a large range of adsorption data [23]. The isotherm is entirely empirical and lacks the theoretical basis for the parameters, and it also provides no physical explanation for the constraints, which makes the interpretation quite challenging. The specific nature of the adsorption energies or interaction on the surface is not described by the model. In addition, at high or low pressures or concentrations, there are changes for the model to fail due to the deviation due to multilayer adsorption or competition for the sites.

2.2.3 Bernauer-Emmett-Teller theory (BET)

The BET theory is a refinement of the Langmuir model where the focus shifts from monolayer to multilayer adsorption, especially near the critical pressure for gas-liquid phase transitions. The BET theory states that absorption would happen in different areas when the critical pressure for liquefaction is low, and this is favoured energetically when compared to the formation of phase-separated droplets. The energy from direct surface binding within the first layer would compensate for the surface exposure costs associated with the liquid droplets [24, 25]. Unlike the phase-separate clusters, the successive layers would have minimal surface tension once the initial layer is formed. The theory does not account for cooperative bonding effects and the surface tension of the multilayer adsorbates, which could lead to wetting or dewetting [25]. **Figure 3** provides an overview of the BET-based binding isotherm, showcasing the different processes or stages involved.

The BET equation is:

$$\frac{p}{v(p_0 - p)} = \frac{1}{v_m c} + \frac{(c-1)p}{v_m c p_o} \tag{5}$$

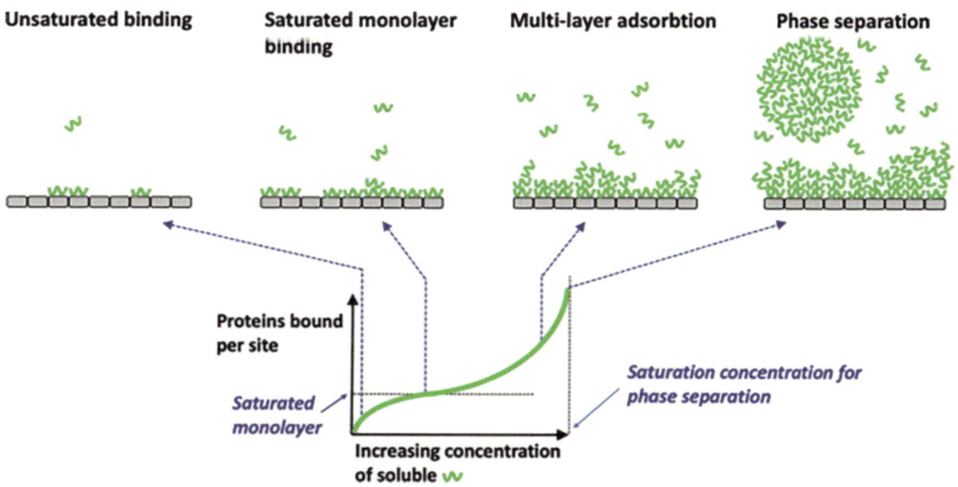

Figure 3.
Hypothetical BET-based binding isotherm [25].

In this equation, p is the equilibrium pressure, while p_0 is the saturation pressure, v is the volume of gas adsorbed and v_m is the monolayer volume of gas adsorbed, with c being the content linked to the adsorption. The theory is widely used for determining the surface area of porous materials [26].

2.3 Thermodynamics of adsorption

The thermodynamics of adsorption would involve the changes in energy, enthalpy, entropy and Gibbs free energy. The adsorption energy quantifies the strength of the interaction that exists between the adsorbate and adsorbent. The exothermic process is found to release energy, which leads to a negative enthalpy change, while physisorption does have lower enthalpy values compared to chemisorption, which is a reflection of weakened interactions. Adsorption generally leads to a decrease in entropy as a result of the restrictions of the adsorbate molecules on the surface, and the change in Gibbs free energy helps determine the spontaneity of adsorption. A negative Gibbs free energy change would indicate a spontaneous process, which is said to happen more likely with the exothermic enthalpy changes and minimal entropy loss.

2.4 Kinetics of adsorption

The kinetics of adoption are used to describe the rate at which the process occurs, along with the mechanisms that are involved. The pseudo-first-order model makes the assumption that the rate of occupation of adsorption sites would be proportional to the number of unoccupied sites [27–29]. This model is said to be used under the physisorption process, where the adsorption makes use of weak van der Wals forces, and the linear form of the pseudo-first-order is represented as:

$$\log(q_e - q_t) = \log q_e - \frac{k_1}{2.303}t \tag{6}$$

In this, q represents the amount of adsorbate that has been adsorbed at time t and equilibrium, followed by k, which is the rate constant of pseudo-first-order adsorption. This model, though, often fails to accurately predict adsorption kinetics over extended periods, especially in situations where the process does not conform to first-order behaviour.

The second-order model was developed by Ho and McKay, and it assumes that the rate of adsorption is proportional to the square of the number of unoccupied sites, which indicates that chemisorption in the rate-limiting step [28, 29]. Based on this, we have the following equation:

$$\frac{t}{q_t} = \frac{1}{k_2 q_e^2} + \frac{t}{q_e} \tag{7}$$

This model is said to be effective in describing the adsorption where there are chemical interactions and often is a better fit for experimental data across the entire adsorption period compared to the previous first-order model. The choice between the models depends on the specific system under study.

2.5 Intraparticle diffusion and rate-controlling steps

Adsorption processes are often finetuned by multiple transport mechanisms like film diffusion, intraparticle diffusion and surface reactions. Film diffusion or external mass transfer would involve the movement of adsorbate molecules from the bulk solution towards the external sauces of the adsorbent, which moves across the boundary layer. The rate of this process would depend on factors like concentration gradient, fluid velocity and overall thickness of the boundary layers [30]. Once the adsorbate molecules reach the external surface, they need to diffuse into the internal pores of the adsorbent in order to access the available sites. This intraparticle diffusion could be divided into macropore and micropore diffusion, which depends on the pore size. The rate of intraparticle diffusion is influenced by the structure, size distribution and tortuosity of the adsorbent material. This is also considered to be the rate-limiting step for porous materials that have complex internal structures [30].

The rate-controlling step is required to optimise the adsorption system, and the Boyd model is commonly used to distinguish between film and intraparticle diffusion control. If a poly of the Boyd function versus time leads to a linear relationship that applies through the origin, then intraparticle diffusion is the rate-controlling step [30]. Deviation from the linearity of intercepts indicates that film diffusion or another process could be the limiting rate. In addition, the Weisz-Prater criterion could be applied to assess the significance of internal diffusion resistance within porous crystals, which would help evaluate the provided diffusion limitation [31, 32]. Thus, understanding and defining the rate-controlling step in the process helps to tailor the system parameters effectively, which improves the overall performance.

3. Techniques for studying adsorption

To optimise the adsorption processes and develop advanced adsorbents, researchers are able to use different techniques to characterise adsorbents, measure adsorption properties and simulate the adsorption mechanisms.

3.1 Characterisation of adsorbents

Characterisation refers to the analysis and understanding of the special and chemical properties of the adsorbent materials that are said to have an impact on their adsorption capacity as well as efficiency. Proper characterisation would involve determining the surface area, pore size distribution, structural integrity and surface functionality. These are the properties that would be given how effective an adsorbent is and can capture and retain adsorbates, which makes characterisation very critical when it comes to material design and application.

3.1.1 Surface and pore size analysis

The surface area and position of the adsorbent are very critical, and the Brunauerr-Emmett-Teller (BET) method is often used for identifying the specific area of porous material, which is through analysing the nitrogen gas adoption isotherms [33, 34]. This also provides the required information on the total surface area, which would be available for adsorption. The Barrett-Joyner Halenda (BJH) is also a method that works well with BET and provides a detailed analysis of the pore size distribution, especially when the material is mesoporous [35, 36]. These techniques are critical in the valuation of materials like activated carbon, metal-organic frameworks (MOF) and zeolites.

3.1.2 Microscopic and spectroscopic techniques

Microscopic techniques like scanning electron microscopy (SEM) and transmission electron microscopy (TEM) help visualise adsorbent surface morphology and pore structures. SEM is used to help provide detailed images of the surface textures, while TEM is found to provide high-resolution imaging of the internal pore arrangements [37, 38]. Other spectroscopic methods like X-ray diffraction help in the identification of the crystalline phases, which helps to better understand the structure stability. Fourier Transform Infrared Spectroscopy (FTIR) is capable of detecting function groups on the surface of the adsorbent, which offers more insight into the chemical nature of the interaction that happens between the adsorbate and adsorbent [38].

3.2 Measurement techniques

To evaluate the adsorption efficiency, different measurement techniques are used, which quantify how much and how fast the adsorbate is captured.

3.2.1 Gravimetric and volumetric methods

Gravimetric methods are used to measure the mass change of adsorbent during the adsorption process, while the volumetric method helps monitor the change in liquid or gas concentration [39, 40]. These techniques help construct the adsorption isotherm, which is then used to illustrate the relationship between the absent and adsorbate concentrations in the equilibrium stage. There is a need to ensure accuracy when both techniques are used, but the extraction of complex and reliable kinetic information from the transient response is more challenging, considering the observed rate could be influenced and controlled by other effects [40].

3.2.2 Calorimetry for heat of adsorption

Calorimetric methods measure the heat that is released or absorbed during the process, which offers thermodynamic insight like enthalpy changes. These measurements are critical in identifying and differentiating between physisorption and chemisorption and measure the number of molecules that interact with the surface, and this provided information could help to understand the binding energy, which helps to describe the adsorbate-adsorbent bonds that are formed theoretically [41, 42].

3.2.3 Chromatography for dynamic studies

Chromatography in an adsorption study is mainly used to help measure dynamic parameters like the adsorption capacity, mass transfer kinetics and breakthrough time. The measure evaluates how the adsorbate would flow through a column that

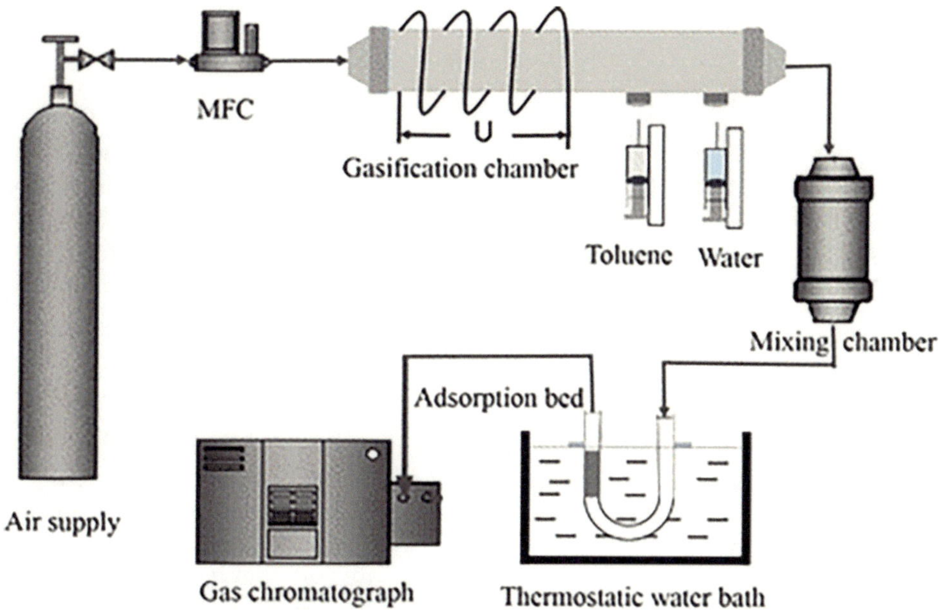

Figure 4.
Dynamic adsorption setup [43].

is packed with an adsorbent when it is put under controlled conditions [43–45]. The output is a breakthrough curve that reflects the concentration of adsorbate as a function of time, and the point indicates when the adsorbents become saturated, while the shape of the curve provides insight into the adsorption efficiency, kinetics and rate-limiting steps. This data are critical in designing and optimising the performance of the adsorbent. **Figure 4** provides an overview of the dynamic adsorption setup that is commonly used in research.

3.3 Modelling and simulation

Computational modelling is said to help in providing atomic level insights into the mechanism, and there are two popular approaches. The first is the molecular dynamics simulation model, which focuses on the dynamic behaviour of the adsorbates on adsorbents, while the Monte Carlo simulations explore equilibrium adsorption properties in different conditions. Both of these techniques are capable of predicting the behaviour and optimising the design based on requirements [46, 47]. Predictive modelling is another approach that uses adsorption isotherms like Langmuir, Freundlich and other isotherms to predict the adsorption activity and behaviour, which would aid in developing efficient systems [48].

4. Advanced adsorbent materials

Adsorbent materials play a critical role in the adoption process and in applications like water purification, gas separation and catalysis. Advanced adsorbents could be classified as emerging and traditional materials, with most of the research focused on enhancing or improving the materials for specific applications.

4.1 Traditional materials

4.1.1 Activated carbon

This is one of the most widely used adsorbents due to the high surface area it possesses, its cost-effectiveness, and its large pore volume. This could be derived from natural sources like wood, coal and coconut shells and find themselves to be sued for different applications like water treatment, air purification and chemical separation. The overall microporous structure is quite effective for adsorbing small molecules like organic contaminants and gases.

4.1.2 Zeolites

These are crystalline structures of aluminosilicates found to possess intricate porous architectures, which make them suitable for several applications involving molecular sieving [49, 50]. The ion exchangeability and the thermal stability are quite favourable for catalytic reactions, water softening and gas separations and they are also found to be more effective in eliminating ammonia or other heavy metals from aster water as they possess cation-exchange properties.

4.1.3 Silica gels

These materials are amorphous, porous materials that have a hydrophilic nature and high thermal stability and are used as a desiccant and also for chromatography, which is attributed to their table pore sizes and ability to adsorb moisture [51, 52]. Their surface could be chemically altered to target specific adsorbates, which would also improve their selectivity for various industrial and environmental applications.

4.2 Emerging materials

4.2.1 Metal-Organic frameworks (MOFs)

They are porous crystalline materials made of metal nodes that are linked to organic compounds and show high surface areas, tunable pore structure and chemical versatility, which makes them ideal for gas storage, carbon dioxide computer and catalysis applications [53–55]. Advances in material design have also allowed for functionalisation to target specific molecules like hydrocarbons or heavy metals.

4.2.2 Covalent-organic frameworks (COFs)

Here, unlike MOFs, covalent bonds are formed between organic compounds that lead to a crystalline material that is lightweight. The high porosity, thermal stability and chemical resistance make them ideal for gas storage, and they have also been used for water purification and energy storage. COFs are also functionalised like MOFs, which further improves their selectivity and incorporation of anime groups, for example, allowing for application in carbon dioxide capture [56–58].

4.2.3 Graphene-based materials

These materials and their derivatives, like graphene oxide (GO) or rGO, have gained a lot of focus in research due to their high surface area, mechanical strength and electronic properties and also their ability to work and improve properties by functionalised with other elements, including MOFs and COFs [59–62]. They are quite effective in adsorbing heavy metals, dyes and organic pollutants, and the conductivity of the material makes them promising for energy storage and catalytic applications.

4.2.4 Bioadsorbents

These are those adsorbent materials derived from renewable sources like agricultural waste or biopolymers that are sustainable alternatives, which makes them attractive in the current situation where there is an increased push for environmentally friendly products. Materials like cellulose have shown that they are capable of removing elements like dyes, heavy metals and other pollutants from wastewater, and their eco-friendly nature and abundance make them very attractive options.

These materials could be mixed and matched to improve the design and modified for specific purposes, which would ensure efficiency and sustainable solutions in different fields.

5. Application of adsorption

Adsorption is really versatile, and depending on the material, it can be used in different sections or for different applications, which are explored in this section.

5.1 Environmental applications

5.1.1 Wastewater treatment

The adsorption process plays a critical role in the purification of water and wastewater treatment, which helps to remove organic pollutants and other elements and make the water pure. Traditional adsorbents like activated carbon and emerging materials like bio adsorbents would make use of different techniques but provide affordable and efficient solutions [63, 64]. Materials like graphene oxide and MOFs have higher selectivity and adsorption capacities, which allow them to be used as tools to remove contaminants such as pharmaceuticals and pesticides from wastewater [65, 66]. The adoption process ensures the recovery of valuable resources like nutrients and natural materials from wastewater, which supports the UN's sustainable goals.

5.1.2 Air pollution control

Activated carbon, for example, is often used for air filters, which helps to remove carbon compounds in air like toluene and other elements that are generated in the industrial exhaust streams. The process would begin with the polluted air passing through the column or filter bed that is packed with the material, and as the molecules diffuse through the air, it would adore into the surface of the activated carbon due to the higher surface area and porous structure. The adsorption process is very effective in removing nearly 95% of the carbon compounds, helps to ensure compliance with environmental regulations, and also improves air quality [67, 68]. This is also scalable and can be used in industries such as petrochemicals, paints and others where emissions are highly prevalent.

5.2 Industrial applications

Zeolites are often used for separating nitrogen from the air in oxygen concentrators, and the process involves feeding compressed air into a column packed with zeolites they adsorb nitrogen due to their molecular sieving property and also affinity towards nitrogen, while oxygen is less adsorbed and exists within the space [69, 70]. After saturation, the zeolite with nitrogen adsorbed would be regenerated by reducing the pressure, which frees the nitrogen atoms, and this process ensures a continuous supply of oxygen for medical and industrial use.

5.3 Catalysis and chemical synthesis

Zeolites are used as catalysts in the production of gasoline through the fluid catalytic cracking (FCC) process, where hydrocarbon feedstock is introduced into a reactor containing zeolite catalysts. The adsorption of hydrocarbon into the surface of the zeolite occurs, and this is where the active acidic sites would catalyse the cracking of the long hydrocarbon chains into smaller, more valuable molecules like gasoline [71, 72]. The products would desorb and be collected while the spent catalyst would be regenerated by a bruin of carbon deposits.

5.4 Biomedical applications

The MOFs are often used in drug delivery for cancer treatment through the use of doxorubicin as the drug, and it begins with loading the molecules into the porous structure of the MOF using adsorption, and then this is administered to the patient, where the acidic environment of cancerous tissues would trigger the release of the medicine [73, 74]. The adsorption and controlled release ensure targeted delivery and help to minimise the side effects while maximising therapeutic efficacy.

This shows that adsorbents have many uses and different adsorbents have different applications, but they can also be created based on requirements.

6. Challenges and future perspectives

6.1 Limitations

Even though existing techniques and materials have made significant advancements, they face several limitations. The first of these is that most of the adsorbents are known to lose their efficiency after repeated use, which is due to structural degradation, fouling or difficulty for the material to regenerate [75, 76]. For example, we have activated carbons that are known to be widely used and require energy-intensive thermal regeneration, which is very costly and hard to use in different applications. It is also reported that magnetic adsorbents are capable of overcoming some of the limitations, as they allow for easy separation using the application of an external magnetic field, which would reduce the overall energy consumption and also minimise the adsorbent loss during the recovery process. That said, to achieve this, it is important to overcome practical application of magnetic adsorbent in wastewater treatment, which would include the stability of materials under different environmental conditions [77].

Another challenge for the use is the high selectivity for the specific adsorbates within complex mixtures, and one example of this would be the difficulty of the conventional adsorbents in differentiating between chemically similar molecules during the gas separation process [77, 78]. This would mean that contaminants could often be added to the mixture, which could cause difficulties or issues with the application. In addition to these, the adsorbents are known to have very low adsorption capacity when it comes to removing contaminants, and this is seen to be more prominent when the application requires high concentration or diverse mixtures [79]. The higher cost of materials like MOF and COF makes them expensive to produce and also limits their scalability for large-scale applications. Often, the synthesis of the materials would require costly reagents along with time-consuming procedures, which makes them impractical for widespread use [80]. The spent adsorbents need to be disposed of properly, and if they are not, then they could lead to environmental pollution and cause more damage.

6.2 Future trends

There has been significant progress made when it comes to improving the absorption capacity by using nanostructured materials like carbon nanotubes, MFs and graphene, as these materials possess very high surface area and also customisable pore structures that allow for efficient binding and removal of diverse contaminants. The unique properties of these materials are not just limited to the surface area, but they help with chemical interaction and confinement effects that would help to boost the adsorption process, and this interaction provides new avenues for capturing and removing contaminants that were previously challenging to address [81, 82]. There has also been a rise in functionalisation techniques which has allowed for the introduction of specific function groups for different applications, and this would allow for the targeted removal of specific contaminants and thus enhance the overall efficiency and selectivity [83]. There have also been innovations that are focused on enhancing the specificity and efficiency of adsorbents using different approaches, and one of the innovations in this would be the Molecularly Imprinted Polymers (MIPs), which are engineered to replicate the molecular recognition that is found within the natural system [84, 85]. Recent research in the field has shown that novel adsorbents that are more resistant to a broad spectrum of physical-chemical stresses are being developed [77]. CNTs and GOs are known to enhance the use of functionalisation and reduction approaches that ensure the presentation of adsorption efficacy.

7. Conclusion

Adsorption is said to be the accumulation of molecules, atoms, or ions on the solid surface, and this is a fundamental process that is different from absorption and is governed by specific molecular interactions, and has different applications. The historical development of science begins with the foundational work of Scheele and Fardat to the advanced theories like BET theory and Langmuir theory that highlight how the process and approach have changed. The research also explored the different mechanisms like physisorption and chemisorption, and this mechanism was found to be influenced by factors like surface area, pore structure and molecular properties, which are critical in optimising the process and adsorbent. The paper also explores the thermodynamic and kinetic analyses that offer insight into the spontaneity, energy changes and rate of adsorption while exploring the isotherms, which helped provide the mathematical framework used to understand the adsorption behaviour under different conditions. Characterisation techniques like BET analysis and measurement models like gravimetric and calorimetric models are also explored, and it shows these methods could be used as a tools to help refine the design and application of an adsorption system.

Over the years, adsorption technologies have made headway in addressing some significant challenges across different industries. Water treatment, air pollution control, gas separation, catalysis and biomedical technologies are among the many applications that demonstrate the efficacy of adsorption – be it with traditional adsorbents like activated carbon and zeolites or with modern MOFs, COFs and graphene-based adsorbents. At the same time, there are many adsorbent technologies that have yet to be fully explored, and hurdles like cost, selectivity and scalability need to be overcome. Further, the regeneration and reusability challenges make the usage of these technologies more difficult on a larger scale. Although progress is already underway in green synthesis, hybrid materials and functionalisation work together to optimise efficiency and drive innovation.

Conflict of interest

The authors declare that there are no conflicts of interest.

Author details

Marwah Ahmed Alsharif
Department of Physics, Faculty of Science, University of Tabuk, Tabouk, Saudi Arabia

*Address all correspondence to: malsharif@ut.edu.sa

References

[1] Alaqarbeh M, Al P, Bin AH. Adsorption phenomena: Definition, mechanisms, and adsorption types: Short review. RHAZES: Green and Applied Chemistry. 2021;**13**:43-51

[2] Siu B, Chowdhury AR, Yan Z, Humphrey SM, Hutter T. Selective adsorption of volatile organic compounds in metal-organic frameworks (MOFs). Coordination Chemistry Reviews. 2023;**485**:215119

[3] Xie Y, Lyu S, Zhang Y, Cai C. Adsorption and degradation of volatile organic compounds by metal–organic frameworks (MOFs): A review. Materials. 2022;**15**(21):7727. DOI: 10.3390/ma15217727

[4] Chai WS, Cheun JY, Kumar PS, Mubashir M, Majeed Z, Banat F, et al. A review on conventional and novel materials towards heavy metal adsorption in wastewater treatment application. Journal of Cleaner Production. 2021;**296**:126589

[5] Sukmana H, Bellahsen N, Pantoja F, Hodur C. Adsorption and coagulation in wastewater treatment - Review. Progress in Agricultural Engineering Sciences. 2021;**17**(1):49-68

[6] Panarin VY, Baum EA, Lanin SN. History of the development of ideas about adsorption in the late 18th and 19th century. Sorbtsionnye i Khromatograficheskie Protsessy. 2019;**19**(2):211-220

[7] Robens E. Some intriguing items in the history of adsorption. Studies in Surface Science and Catalysis. 1994;**87**(C):109-118

[8] Rakshit AK, Moulik SP. Centenary of langmuir isotherm. Journal of Surface Science and Technology. 2019;**35**(1-2):1-16

[9] Carrete J, García M, Rodríguez JR, Cabeza O, Varela LM. Theoretical model for moisture adsorption on ionic liquids: A modified Brunauer-Emmet-Teller isotherm approach. Fluid Phase Equilibria. 2011;**301**(1):118-122

[10] Alhamami M, Doan H, Cheng CH. A review on breathing behaviors of metal-organic-frameworks (MOFs) for gas adsorption. Materials. 2014;**7**(4):3198-3250

[11] Sağlam S, Türk FN, Arslanoğlu H. Use and applications of metal-organic frameworks (MOF) in dye adsorption: Review. Journal of Environmental Chemical Engineering. 2023;**11**:110568

[12] Zhang T, Zhang G, Chen L. 2D conjugated covalent organic frameworks: Defined synthesis and tailor-made functions. Accounts of Chemical Research. 2022;**55**(6):795-808

[13] Abdelhamid HN. An introductory review on advanced multifunctional materials. Heliyon. 2023;**9**:e18060. DOI: 10.1016/j.heliyon.2023.e18060

[14] Mahmood Aljamali N, Abdul Baqi Aldujaili D, Obaid AI. Physical and chemical adsorption and its applications. International Journal. 2021;**7**(2):1-9

[15] Nikolic K, Agababa D. Design and QSAR study of analogs of γ-tocotrienol with enhanced antiproliferative activity against human breast cancer cells. Journal of Molecular Graphics & Modelling. 2009;**27**(7):982-991

[16] Nemes Z, Dietz R, Lüth JB, Gomba S, Hackenthal E, Gross F. The

pharmacological relevance of vital staining with neutral red. Experientia. 1979;**35**(11):1474-1475

[17] Kennedy KK, Maseka KJ, Mbulo M. Selected adsorbents for removal of contaminants from wastewater: Towards engineering clay minerals. Open Journal of Applied Sciences. 2018;**08**(08):419-431

[18] Shoemaker M, Cohen I, Campbell M. Reduction of MTT by aqueous herbal extracts in the absence of cells. Journal of Ethnopharmacology. 2004;**93**(2-3):381-384

[19] Bernal V, Giraldo L, Moreno-Piraján JC. Insight into adsorbate–adsorbent interactions between aromatic pharmaceutical compounds and activated carbon: Equilibrium isotherms and thermodynamic analysis. Adsorption. 2020;**26**(2):349-359

[20] Kokalj A. On the use of the Langmuir and other adsorption isotherms in corrosion inhibition. Corrosion Science. 2023;**217**:110993

[21] Liu Y. Some consideration on the Langmuir isotherm equation. Colloids and Surfaces A: Physicochemical and Engineering Aspects. 2006;**274**(1-3):65-71

[22] Ehiomogue P, Ahuchaogu II, Ahaneku IE. Review of adsorption isotherms models. Acta Technica Corviniensis. 2022;**14**(4):57-62

[23] Kalam S, Abu-Khamsin SA, Kamal MS, Patil S. Surfactant adsorption isotherms: A review. ACS Omega. 2021;**6**(52):35753-35768

[24] Aps PA. Brunauer, Emmett and Teller (BET) Theory. Vol. 300. Available from: http://Particle.Dk/Methods-Analytical-Laboratory/Surface-Area-Bet/Surface-Area-Bet-Theory/2014

[25] Mitchison TJ. Beyond langmuir: Surface-bound macromolecule condensates. Molecular Biology of the Cell. 2020;**31**(4):482-488

[26] Brunauer S, Emmett PH, Teller E. Adsorption of gases in multimolecular layers. Journal of the American Chemical Society. 1938;**60**(2):309-319

[27] Revellame ED, Fortela DL, Sharp W, Hernandez R, Zappi ME. Adsorption kinetic modeling using pseudo-first order and pseudo-second order rate laws: A review. Cleaner Engineering and Technology. 2020;**1**:100007

[28] Khamizov RK. A pseudo-second order kinetic equation for sorption processes. Russian Journal of Physical Chemistry A. 2020;**94**(1):145-148

[29] Lima ÉC, Adebayo MA, Machado FM. Kinetic and equilibrium models of adsorption. In: Bergmann C, Machado F, editors. Carbon Nanomaterials as Adsorbents for Environmental and Biological Applications. Carbon Nanostructures. Cham: Springer; 2015. pp. 33-69. DOI: 10.1007/978-3-319-18875-1_3

[30] Viegas RMC, Campinas M, Costa H, Rosa MJ. How do the HSDM and Boyd's model compare for estimating intraparticle diffusion coefficients in adsorption processes. Adsorption. 2014;**20**(5-6):737-746

[31] Vishwakarma R, Mannepalli LK, Rathod V. Kinetics of Henry reaction catalyzed by fluorapatite. Chemical Engineering Research and Design. 2022;**181**:495-504

[32] García JR, Bidabehere CM, Sedran U. Diffusion controlled LHHW kinetics.

Simultaneous determination of chemical kinetic and equilibrium adsorption constants by using the Weisz-prater approach. Chemical Engineering Science. 2017;**172**:18-25

[33] Ta DN, Nguyen HKD, Trinh BX, Le QTN, Ta HN, Nguyen HT. Preparation of nano-ZIF-8 in methanol with high yield. Canadian Journal of Chemical Engineering. 2018;**96**(7):1575-1581

[34] Virtanen T, Rudolph G, Lopatina A, Al-Rudainy B, Schagerlöf H, Puro L, et al. Analysis of membrane fouling by Brunauer-Emmet-Teller nitrogen adsorption/desorption technique. Scientific Reports. 2020;**10**(1):1-11

[35] Nguyen TTD, Choi HN, Ahemad MJ, Van Dao D, Lee IH, Yu YT. Hydrothermal synthesis of In_2O_3 nanocubes for highly responsive and selective ethanol gas sensing. Journal of Alloys and Compounds. 2020;**820**:157280. ISSN 0925-8388

[36] Kimura M, Qi ZD, Isogai A. Analysis of mesopore structures in wood cell walls and pulp fibers by nitrogen adsorption method. Nordic Pulp & Paper Research Journal. 2016;**31**(2):244-249

[37] Medhat A, El-Maghrabi HH, Abdelghany A, Abdel Menem NM, Raynaud P, Moustafa YM, et al. Efficiently activated carbons from corn cob for methylene blue adsorption. Applied Surface Science Advances. 2021;**3**:100063

[38] Rahman A, Haque MA, Ghosh S, Shinu P, Attimarad M, Kobayashi G. Modified shrimp-based chitosan as an emerging adsorbent removing heavy metals (chromium, nickel, arsenic, and cobalt) from polluted water. Sustainability (Switzerland). 2023;**15**(3):2-20

[39] Gómez-Gualdrón DA, Wang TC, García-Holley P, Sawelewa RM, Argueta E, Snurr RQ, et al. Understanding volumetric and gravimetric hydrogen adsorption trade-off in metal-organic frameworks. ACS Applied Materials & Interfaces. 2017;**9**(39):33419-33428

[40] Wang JY, Mangano E, Brandani S, Ruthven DM. A review of common practices in gravimetric and volumetric adsorption kinetic experiments. Adsorption. 2021;**27**:637-655

[41] Fubini B. Adsorption calorimetry in surface chemistry. Thermochimica Acta. 1988;**135**(C):21-29

[42] Garcia-Cuello V, Moreno-Piraján JC, Giraldo-Gutiérrez L, Sapag K, Zgrablich G. Determination of differential enthalpy and isotherm by adsorption calorimetry. Research Letters in Physical Chemistry. 2008;**2008**:1-4

[43] Hu L, Cheng W, Zhang W, Wu F, Peng S, Li J. Monolithic bamboo-based activated carbons for dynamic adsorption of toluene. Journal of Porous Materials. 2017;**24**(2):553-563

[44] Kondor A, Santmarti A, Mautner A, Williams D, Bismarck A, Lee KY. On the BET surface area of nanocellulose determined using volumetric, gravimetric and chromatographic adsorption methods. Frontiers in Chemical Engineering. 2021;**3**:736473

[45] Chen KH, Lai YR, Hanh NTD, Wang SSS, Chang YK. Breakthrough curve modeling and analysis for lysozyme adsorption by Tris(hydroxymethyl)aminomethane affinity nanofiber membrane. Membranes (Basel). 2023;**13**(9):894

[46] Hamed O, Qaisi M, Abushqair I, Berisha A, Dagdag O, Janem A, et al.

Cellulose powder functionalized with phenyl biguanide: Synthesis, cross-linking, metal adsorption, and molecular docking. BioResources. 2021;**16**:3127-3142.19

[47] Lebkiri I, Abbou B, Hsissou R, Safi Z, Sadiku M, Berisha A, et al. Investigation of the anionic polyacrylamide as a potential adsorbent of crystal violet dye from aqueous solution: Equilibrium, kinetic, thermodynamic, DFT, MC and MD approaches. Journal of Molecular Liquids. 2023;**372**:131308

[48] Igwegbe CA et al. Modeling of adsorption of methylene blue dye on Ho-CaWO4 nanoparticles using response surface methodology (RSM) and artificial neural network (ANN) techniques. 2019. DOI: 10.1016/j.mex.2019.07.016

[49] Campanile A, Liguori B, Ferone C, Caputo D, Aprea P. Zeolite-based monoliths for water softening by ion exchange/precipitation process. Scientific Reports. 2022;**12**(1):16207

[50] Xu C, Wei W, He Y. Enhanced hydrogen separation performance of linde type-a zeolite molecular sieving membrane by cesium ion exchange. Materials Letters. 2022;**324**:132536

[51] Ge TS, Qi D, Dai YJ, Wang RZ. Experimental testing on contaminant and moisture removal performance of silica gel desiccant wheel. Energy and Buildings. 2018;**176**:135-144

[52] Pramuang S, Exell RHB. The regeneration of silica gel desiccant by air from a solar heater with a compound parabolic concentrator. Renewable Energy. 2007;**32**(1):173-182

[53] Felix Sahayaraj A, Joy Prabu H, Maniraj J, Kannan M, Bharathi M, Diwahar P, et al. Metal–organic frameworks (MOFs): The next generation of materials for catalysis, gas storage, and separation. Journal of Inorganic and Organometallic Polymers and Materials. 2023;**33**:355-375

[54] Kim HR, Yoon TU, Kim SI, An J, Bae YS, Lee CY. Beyond pristine MOFs: Carbon dioxide capture by metal-organic frameworks (MOFs)-derived porous carbon materials. RSC Advances. 2017;**7**(3):1266-1270

[55] Xiao C, Tian J, Chen Q, Hong M. Water-stable metal-organic frameworks (MOFs): Rational construction and carbon dioxide capture. Chemical Science. 2024;**15**:225-240

[56] Zhou L, Li S, Chen L, Li Q, Lu C, Tan L, et al. MOFs and COFs based pervaporation membranes for alcohols/water separation: A review. Separation and Purification Technology. 2024;**330**:124934

[57] Zhang H, Geng Y, Huang J, Wang Z, Du K, Li H. Charge and mass transport mechanisms in two-dimensional covalent organic frameworks (2D COFs) for electrochemical energy storage devices. Energy and Environmental Science. 2023;**16**:889-951

[58] Wu MX, Yang YW. Applications of covalent organic frameworks (COFs): From gas storage and separation to drug delivery. Chinese Chemical Letters. 2017;**28**:1135-1143

[59] Catherine HN, Tan KH, Shih Y, Ran D, Manu B, Yuan DJ. Surface interaction of tetrabromobisphenol a, bisphenol a and phenol with graphene-based materials in water: Adsorption mechanism and thermodynamic effects. Journal of Hazardous Materials Advances. 2023;**9**:100227

[60] da Silva Alves DC, Healy B, Pinto LA d A, TRS C, Breslin CB. Recent developments in chitosan-based adsorbents for the removal of pollutants from aqueous environments. Molecules. 2021;**19**:5789

[61] Xu L, Xiao G, Chen C, Li R, Mai Y, Sun G, et al. Superhydrophobic and superoleophilic graphene aerogel prepared by facile chemical reduction. Journal of Materials Chemistry A. 2015;**3**(14):7498-7504

[62] Weng Z, Su Y, Wang DW, Li F, Du J, Cheng HM. Graphene-cellulose paper flexible supercapacitors. Advanced Energy Materials. 2011;**1**(5):917-922

[63] Karić N, Maia AS, Teodorović A, Atanasova N, Langergraber G, Crini G, et al. Bio-waste valorisation: Agricultural wastes as biosorbents for removal of (in)organic pollutants in wastewater treatment. Chemical Engineering Journal Advances. 2022;**9**:100222

[64] Lubis RAF, Nasution HI, Zubir M. Production of activated carbon from natural sources for water purification. Indonesian Journal of Chemical Science and Technology (IJCST). 2020;**3**(2):47 52

[65] Wang S, Li X, Liu Y, Zhang C, Tan X, Zeng G, et al. Nitrogen-containing amino compounds functionalized graphene oxide: Synthesis, characterization and application for the removal of pollutants from wastewater: A review. Journal of Hazardous Materials. 2018;**342**:177-191

[66] Su G, Ong HC, Ibrahim S, Fattah IMR, Mofijur M, Chong CT. Valorisation of medical waste through pyrolysis for a cleaner environment: Progress and challenges. Environmental Pollution. 2021;**279**:116934

[67] Sircar S, Golden TC, Rao MB. Activated carbon for gas separation and storage. Carbon N Y. 1996;**34**(1):1-12

[68] Chuah CY, Laziz AM. Recent Progress in synthesis and application of activated carbon for CO_2 capture. C-Journal of Carbon Research. 2022;**8**(2):25

[69] Detho A, Memon AH, Alali AF, Almohana AI, Almojil SF, Memon AA, et al. Ammoniacal nitrogen, chemical oxygen demand, and color reduction in rubber processing industry effluent using zeolite. Desalination Water Treat. 2022;**270**:112-121

[70] Jayaraman A, Yang RT, Cho SH, Bhat TSG, Choudary VN. Adsorption of nitrogen, oxygen and argon on Na-CeX zeolites. Adsorption. 2002;**8**(4):325-333

[71] Degnan TF. Applications of zeolites in petroleum refining. Topics in Catalysis. 2000;**13**(4):349-356

[72] Muir B, Bajda T. Organically modified zeolites in petroleum compounds spill cleanup - Production, efficiency, utilization. Fuel Processing Technology. 2016;**149**:347-354

[73] Lv L, Fu Z, You Q, Xiao W, Wang H, Wang C, et al. Enhanced photodynamic therapy through multienzyme-like MOF for cancer treatment. Frontiers in Bioengineering and Biotechnology. 2023;**11**:1338257

[74] Alijani H, Noori A, Faridi N, Bathaie SZ, Mousavi MF. Aptamer-functionalized Fe_3O_4@MOF nanocarrier for targeted drug delivery and fluorescence imaging of the triple-negative MDA-MB-231 breast cancer cells. Journal of Solid State Chemistry. 2020;**292**:121680

[75] Santos DH d S, Xiao Y, Chaukura N, Hill JM, Selvasembian R, CLPS Z, et al.

Regeneration of dye-saturated activated carbon through advanced oxidative processes: A review. Heliyon. 2022;**8**(8):e10205. ISSN 2405-8440

[76] Nieto-Sandoval J, El MF, Munoz M, Lopez-Arago N, de Pedro ZM, Casas JA. In-situ regeneration of a novel Fe_3O_4/GAC adsorbent for micropollutants removal in a continuous fixed-bed. Journal of Hazardous Materials Advances. 2023;**10**:100287

[77] Satyam S, Patra S. Innovations and challenges in adsorption-based wastewater remediation: A comprehensive review. Heliyon. 2024;**10**(9):e29573

[78] Raji Z, Karim A, Karam A, Khalloufi S. Adsorption of heavy metals: Mechanisms, kinetics, and applications of various adsorbents in wastewater remediation—A review. Waste. 2023;**1**(3):236-258

[79] Younas F, Mustafa A, Farooqi ZUR, Wang X, Younas S, Mohy-Ud-din W, et al. Current and emerging adsorbent technologies for wastewater treatment: Trends, limitations, and environmental implications. Water (Switzerland). 2021;**13**:273

[80] Ren J, Dyosiba X, Musyoka NM, Langmi HW, Mathe M, Liao S. Review on the current practices and efforts towards pilot-scale production of metal-organic frameworks (MOFs). Coordination Chemistry Reviews. 2017;**352**:187-219

[81] Rodríguez C, Tapia C, Leiva-Aravena E, Leiva E. Graphene oxide–Zno nanocomposites for removal of aluminum and copper ions from acid mine drainage wastewater. International Journal of Environmental Research and Public Health. 2020;**17**(18):6708

[82] Ali ME, Hoque ME, Safdar Hossain SK, Biswas MC. Nanoadsorbents for wastewater treatment: Next generation biotechnological solution. International journal of Environmental Science and Technology. 2020;**17**:4095-4118

[83] Rathi BS, Kumar PS. Application of adsorption process for effective removal of emerging contaminants from water and wastewater. Environmental Pollution. 2021;**280**:116990

[84] Kang MS, Cho E, Choi HE, Amri C, Lee JH, Kim KS. Molecularly imprinted polymers (MIPs): Emerging biomaterials for cancer theragnostic applications. Biomaterials Research. 2023;**27**:45. DOI: 10.1186/s40824-023-00388-5

[85] Li G, Row KH. Recent applications of molecularly imprinted polymers (MIPs) on micro-extraction techniques. Separation and Purification Reviews. 2018;**47**(1):1-23

Chapter 2

Selective Adsorption of Chiral Nanomaterials: Mechanisms and Biomedical Applications

Yana Andreevna Gromova and Tatyana I. Shabatina

Abstract

In recent years, the selective adsorption of chiral substances has attracted considerable research attention due to its potential applications in biomedicine. This process involves the use of materials capable of selectively adsorbing one of the chiral isomers, which opens the possibility to improve drugs and create new diagnostic and therapeutic methods. Current scientific research is generally focused on the development and improvement of materials capable of selective adsorption of chiral substances, which can facilitate more precise delivery of drugs into the body, minimizing side effects and improving treatment efficacy. Thus, selective adsorption of chiral molecules represents a promising direction for the future of biomedical technology and medicinal chemistry. In this overview, chapter examples of efficient selective adsorbents for chiral substances, adsorption mechanisms and current biomedical applications of such materials will be discussed.

Keywords: adsorption, chirality, chiral substance, nanoparticles, supramolecular aggregates, biomedicine

1. Introduction

Chirality is one of the fundamental properties of molecules that play a key role in biology and medicine. Many biomolecules such as amino acids, proteins and nucleic acids, as well as synthetic molecules such as antibiotics are chiral. The ability to distinguish, identify and isolate enantiomers is crucial for biological processes and therapeutic effects of drugs [1–5].

Chirality is an essential property of an organism and plays a key role in the regulation of normal physiological processes. The basic structure of all organisms consists of chiral molecules, and biological systems are highly selective to chiral biomolecules. Chirality can be imparted to any material having both natural and artificial origin in the process of asymmetric assembly or grafting of molecular chiral groups or linkers [6].

In recent decades, research in this area has mainly focused on the study of chirality of organic molecules and their optical properties [5]. However, inorganic chiral materials, as well as hybrid materials combining both organic and inorganic fragments, are actively gaining popularity [7–10]. Such materials have chiral crystallographic lattice, possess obvious chiral morphology and can be ideal building blocks like natural chiral objects. However, there are still questions related to understanding the phenomenon of chirality and its occurrence in metallic and semiconductor materials. This question is acute for researchers in contrast to the clear definition of chirality in organic systems.

Chiral inorganic nanomaterials (NMs), a new category of biomaterials, are also becoming more popular. Chiral inorganic NMs have unique physicochemical properties that distinguish them from conventional NMs. In recent years, more and more scientific research has been devoted to the study of various chiral NMs and their physicochemical properties, which directly determine their practical applications in various fields: medicine, theranostics, development of sensor materials, catalytic and adsorption materials. The current review is devoted to different types of chiral NMs and the study of chirality mechanisms, which is an urgent task in today's science. The adsorption patterns in the creation of chiral NMs are also discussed, and their potential applications in the field of biomedicine are discussed [11].

This overview chapter includes an introduction (1), three parts of the main text (2–4) and a conclusion (5). In the second part of the review, we will present the modern classification of chiral materials and talk about the types of chirality. In the third part of the review, we present examples of works focused on the production of chiral NMs and discuss the adsorption process and its mechanism as a key step in the creation of such materials. In the fourth part of the review, we will present the adsorption properties of a hybrid system based on silver nanoparticles and cholesterol derivatives and discuss the possible applications of such a system in biomedical applications.

2. Chiral substances: Classifications and types of chirality

Chiral materials can be either organic or inorganic in nature. As mentioned earlier, many modern chiral materials combine both organic and inorganic moiety. This approach allows the creation of new materials, thereby enhancing one or another physical and chemical properties. Synthetic possibility based on control in the creation of new materials expands the range of possible biomedical applications ranging from diagnostics to therapy in the treatment of various diseases [12–27].

Classification of organic chiral substances is given schematically in **Figure 1**. Substances can be of natural, semisynthetic and synthetic origin. They are created by incorporating molecular chirality into natural molecules (proteins, oligosaccharides, polysaccharides, antibiotics and alakaloids) through asymmetric assembly or grafting of molecular chiral groups or linkers [13]. Such materials produce different optical signals when examined by circular dichroism spectroscopy.

Inorganic chiral materials, namely chiral inorganic NMs, are of increasing interest [3, 14, 15, 25]. The transition to nanoobjects is due to their high demand in very different fields of nanotechnology, biotechnology and medicine. Chiral inorganic NMs have unique physical and chemical properties that distinguish them from conventional NMs. For example, chiral inorganic nanoparticles or nanocomposites have distinct dimensional and structural characteristics that distinguish them from small molecules and traditional materials. Chiral inorganic NMs have a number of advantages over achiral materials: wide range of size distribution, high values of g-factor and the presence of symmetric mirror conformations. Specific physicochemical properties

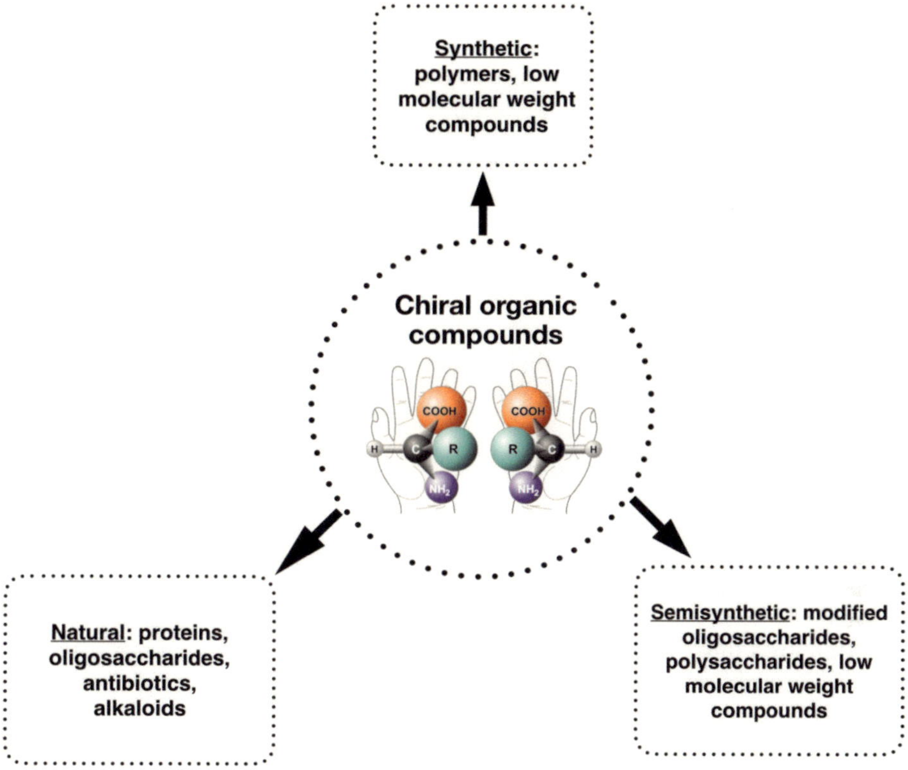

Figure 1.
Classification of organic chiral substances.

of chiral NMs determine their success in biomedicine: screening, diagnostics and therapy, as well as drug delivery.

Chiral inorganic NMs can be separated into different types schematically represented in **Figure 2**:

- chiral metallic nanoparticles/metal oxides,
- chiral non-metallic nanoparticles,

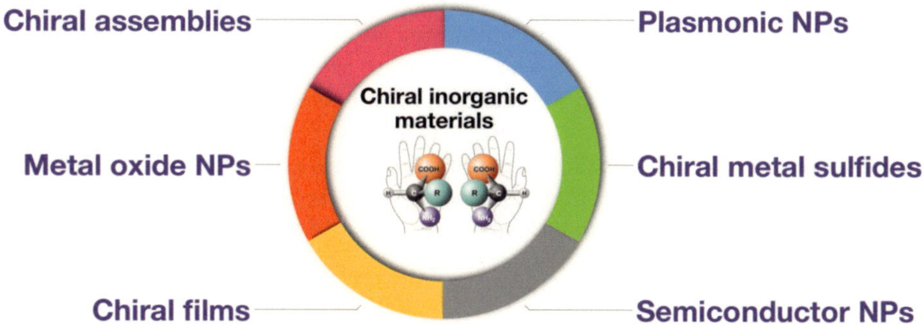

Figure 2.
Classification of inorganic chiral substances.

- chiral semiconductor nanomaterials,
- chiral nanofilms,
- chiral nanofilms, and
- chiral magnetic nanoparticles

Classical inorganic NMs have several disadvantages compared to chiral nanomaterials: biochemical stability, high toxicity and the possibility of metastatic recurrence [16–19]. All this limits their application in biomedicine, for example, as directed drug delivery systems. Giving NMs chirality eliminates numerous drawbacks and opens new opportunities for biomedical applications.

- *First and foremost,* chiral inorganic NMs increase therapeutic efficacy and have high stability in the bloodstream.
- *Second*, chiral inorganic NMs have good biocompatibility and can effectively reduce cytotoxicity.
- *Third*, chiral inorganic NMs absorb left- and right-polarized light in completely different ways, which makes it possible to increase the accuracy of their activation in the lesion focus.

The above properties have an evidentiary basis, chiral NMs, including metallic nanoparticles and nanocomposites, are actively used in medicine in therapy, treatment and prevention of a wide range of diseases, including cancer [19].

Among chiral inorganic nanostructures, the main types are chiral *nanoparticles* (*NPs*) and chiral inorganic nanostructures, which usually represent a "core-shell" model where the core is the surface of the NPs, and the shell is the ligand. Most often, the occurrence of chirality can be related to the nature of the stabilizing chiral ligand, as well as the shape of the inorganic core (sphere, cube, star, etc.) or chiral rearrangement of the NPs surface because of stabilization by the ligand. Also, chirality can be related to the initial asymmetry of the NP. Now, there are four main causes of chirality (**Figure 3**) [2].

The first mechanism of chirality arises from the asymmetry of the inorganic core within nanoparticles (NPs). Typically, particularly in "bottom-up" synthesized NPs, the inorganic core is regarded as a single structural unit. Even when the NPs possess a perfectly spherical geometry (with inorganic cores forming ideal crystals) and their surface ligands are uniformly distributed, the material can still demonstrate chirality. This phenomenon occurs because external electric or magnetic fields can induce asymmetry in the electron density of the inorganic cores through polarization effects. The chiral structure of the inorganic nucleus is favored as chiral lattice distortions are more thermodynamically stable. For instance, an achiral tetrahedron has only one possible atomic configuration for its vertices, whereas a chiral tetrahedron features four distinct vertices that can be arranged in 24 unique ways [2]. Compared to the achiral tetrahedron, the asymmetric tetrahedron's atomic arrangement increases entropy by approximately 8 kJ/mol, making it energetically more favorable.

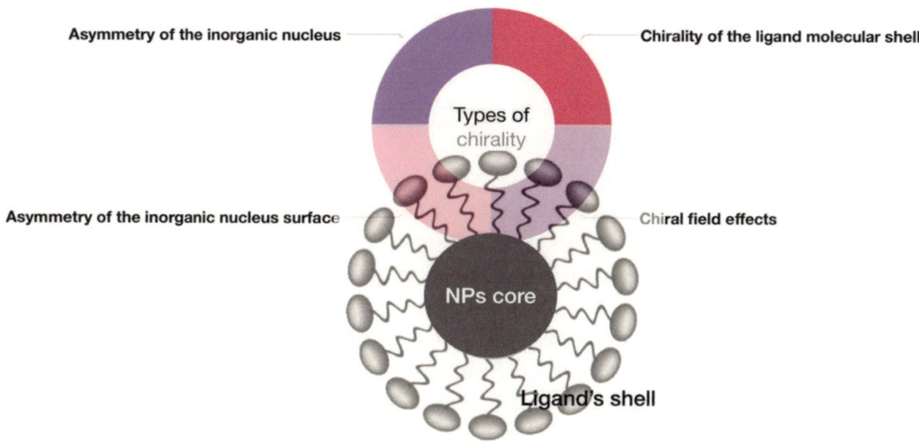

Figure 3.
Types of chirality.

The second type of chirality is related to the asymmetry of the surface of the inorganic core of the NP. In this case, the nucleus itself is achiral, while the surface of inorganic materials is chiral. This type of chirality is the result of deformation by substitution of surface atoms of the nucleus with ligand-stabilizer functional groups. The ligand itself can be chiral or achiral. In this case, only the surface energy of the material changes; the nuclear component is stationary. Many ligands that lead to lattice deformation and the appearance of chirality have been described nowadays. The most pronounced deformation effect is observed with ligands with two functional groups. For example, cysteine and tartaric acid containing -COOH and -SH groups. When chiral molecules are sorbed on the surface of metal particles, the atoms on the surface adopt a special arrangement that breaks all local planes of symmetry.

The third type of chirality is the nature of the ligand and it is shell. In this case, the nucleus and surface of the NPs are achiral, and the molecular shell of the ligand is chiral. Also, the chirality can be caused by achiral ligands (there is a symmetry breaking at stabilization of ligands of surface atoms). From the practical point of view, the "creation" of such chirality is the simplest and most variable practical approach. It is based on the use of different types of nanoparticles (metallic, non-metallic) and varying the nature of the ligand which makes it possible to obtain new materials that are actively used as adsorption and separation materials.

The fourth type of chirality is the chirality of NPs due to chiral field effects. This is the rarest type of chirality, and it takes place only in strongly polarized inorganic materials. In this case, the distribution of atoms in the NPs nucleus is achiral, but the distribution of electrons can be affected by ligands or other influences, which leads to asymmetric polarization. Such chirality is difficult to fix during the experiment: asymmetric polarization of electrons leads to the fact that NPs, especially plasmonic (Ag and Au), when interacting with light give asymmetric absorption and often show strong circular dichroism (CD) signals.

In recent years, hybrid materials related to *metal-organic materials* (MOMs), namely *metal-organic frameworks* (MOFs), have been gaining popularity. In [20], the author and his co-authors proposed to classify all solid porous materials into three types:

- organic (polymers),
- inorganic (zeolites), and
- metal-organic frameworks (MOFs).

MOFs are modern hybrid materials that include organic molecules acting as linkers and metal ions/clusters. The formation of MOFs occurs due to various intermolecular interactions (hydrogen, coordination bonds, electrostatic attraction), which leads to the formation of one-dimensional, two-dimensional or three-dimensional structures [7]. Such materials have several obvious advantages: chemical stability, large and controlled porosity and high specific surface area. All this makes MOFs a promising and versatile class of materials with wide practical applications. The use of different organic linkers for coordination with metal ions has facilitated the synthesis of innovative chiral materials in terms of potential applications. According to the pre-designed target molecule, it is also possible to control the hydrophobicity or hydrophilicity of the material (**Figure 4**).

Crystalline MOFs are promising adsorption and separation materials and have therefore been successfully utilized in chiral catalysis and chromatographic separation techniques [22, 23]. It is possible to regulate the orientation of two enantiomers in a confined microenvironment by incorporating several chiral recognition sites into such porous materials. This will facilitate enantiospecific interactions, enantiomer identification and efficient stereocontrol. Compared to zeolites, MOFs have become an ideal basis for the development of new porous materials, which opens a wide range of applications in various research fields. They have higher porosity than conventional inorganic and organic porous materials (activated carbons and zeolites), structural and chemical diversity, functional variability and choice of size and shape [20].

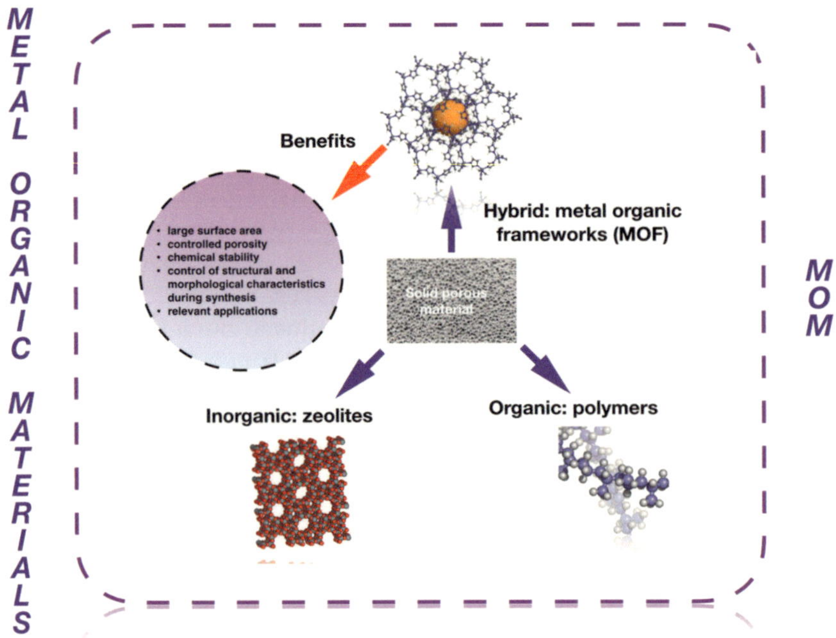

Figure 4.
Types of solid porous materials.

3. Chiral substances: Adsorption of chiral nanomaterials

In the first section, we highlighted the chirality effect and its practical significance, discussed possible sources of chirality and gave modern classifications of chiral substances. In this section, we present some works reflecting the importance of the adsorption process in the creation of new chiral nanomaterials, mainly for biomedical applications.

One of the most common porous materials is silica [28–32]. This material has a high specific surface area, is thermally and mechanically stable and can be easily modified due to the developed surface area due to the multitude of silanol groups. The authors [28] studied the process of oral adsorption of non-steroidal anti-inflammatory drugs in the gastrointestinal mucosa using mesoporous chiral AT-R-CMSN silica and mesoporous achiral silica S-MSN as drug carriers. Poor adhesion of nanocarriers to the intestinal mucosa is one of the main causes of failure in oral drug delivery. Mesoporous silica nanoparticles with a chiral geometric structure, AT-R-CMSN, were engineered to enhance nanoscale surface roughness and utilized as carriers for poorly soluble drugs like nimesulide and ibuprofen. The process of adsorption plays a critical role in the development of these materials. Nitrogen adsorption analysis revealed that drug-free S-MSN and AT-R-CMSN exhibit a uniform mesoporous architecture, as indicated by type IV isotherms with distinct condensation steps. Additionally, the nanoparticles demonstrated consistent pore sizes (2.49 and < 2.17 nm, respectively) with a narrow size distribution. Their SBET values were 821.5 cm^2/g for AT-R-CMSN and 523.5 cm^2/g for S-MSN, highlighting their capability to store or adsorb a wide range of substances. Both materials displayed excellent thermal stability, with minimal weight loss (<5%) attributed to adsorbed moisture. Consequently, the drug-loaded forms, nimesulide-AT-R-CMSN and ibuprofen-AT-R-CMSN, exhibited enhanced anti-inflammatory efficacy and significantly improved bioavailability (705.95 and 444.42%, respectively) compared to other formulations. The chiral architecture of AT-R-CMSN effectively addressed challenges associated with the stability, solubility, and permeability of insoluble drugs, paving the way for the design of advanced drug delivery systems based on mesoporous chiral silica gels [28].

Parallel research, as noted in [31], focused on addressing clinical demands for improved oral bioavailability and reduced side effects of insoluble drugs. Aminated mesoporous silica gel (L/D-BPEIN MSX), featuring a chiral surface topology, was synthesized using a rapid and cost-effective method within 2 minutes, serving as a carrier for indomethacin, a non-steroidal anti-inflammatory drug. The synthesis involved the self-assembly of L/D-tartaric acid (L/D-TA) with branched polyethylenimine (BPEI) to impart chirality, synergistically enhancing silica deposition. Simultaneously, 3-aminopropyltriethoxysilane (APTES) was polycondensed with silica precursors to form an amino-functionalized mesostructure. Nitrogen adsorption analysis of BPEIN-MSX and L/D-BPEIN-MSX showed type IV isotherms with prominent capillary condensation steps, confirming mesoporosity. The hysteresis loop observed for BPEIN-MSX at higher relative pressures (0.6–0.8) indicated a looser pore arrangement due to electrostatic repulsion caused by APTES amino groups. In contrast, L/D-BPEIN-MSX exhibited smaller pore sizes due to reduced electrostatic repulsion and synergistic effects on silica deposition from tartaric acid. Drug loading capacities were determined to be 31.4% for BPEIN-MS and 26% for L/D-BPEIN-MSX. Notably, L/D-BPEIN-MSX significantly enhanced the release rate of indomethacin *via* various controlled mechanisms. Additionally, D-BPEIN-MSX demonstrated chiral responsiveness in the microenvironment, while L-BPEIN-MSX favored bioprocesses like contact and adhesion due to its preferential chiral surface. These carriers improved indomethacin's bioavailability by eight to nine times, delivering effective anti-inflammatory action without gastrointestinal irritation [31].

It is well known that the adsorption capacity of silica can be improved by modifying its surface with various compounds containing NH_2- and SH-functional groups [33–36]. Such functionalized silica is often used as a carrier for binding to metal nanoparticles, including antibacterial AgNPs. Silver shows a higher affinity for thiol groups, so the bond between them is formed stronger than with NH_2-groups. For this reason, AgNPs immobilized on the surface of SH-silica usually have a narrower size distribution. In addition, the surface of thiolated silica can "graft" more silver. In work [33], SH-functionalized silicon granules were prepared for the adsorption of AgNPs from their solution. It is known that AgNPs have a high affinity toward S-containing compounds. Therefore, the adsorption mechanism is simple and easy to explain: the formation of complexes between Ag + ions and SH-groups on the carrier surface during adsorption from aqueous solution. The amount of silver adsorbed by the surface depends not only on the amount and density of S-atoms on the surface of the modified silica but also on the accessibility of Ag + ions to these functional groups. The adsorption behavior of AgNPs on the surface follows the Langmuir model. The amount of adsorbed AgNPs was 75.8 mg/g [33].

Chiral nanomaterials can be synthesized by altering the composition of AgNP-based composites [35], where silica acts as the carrier, silver nanoparticles serve as the core, and ligand stabilizers form the shell. The ligand's nature is a primary source of chirality, as detailed in Section 1 [36–41]. Beyond silica-based materials, carbon nanomaterials (CNMs) are emerging as promising candidates for chiral separation [42–48]. CNMs possess a high specific surface area, good biocompatibility, delocalized π-electron systems, and abundant oxygen-containing groups, all of which enhance chiral separation efficiency. For instance, graphene oxide (GO), a notable CNM, has been extensively explored for its hybrid materials in adsorption and separation applications [42]. Researchers [43] developed a hybrid material based on spiral silicon nanotubes (GO-SNTs) encapsulated in GO, where changes in nanotube helicity allowed tuning of enantioselectivity. Left-handed GO-SNTs selectively bound L-form amino acid derivatives, while right-handed GO-SNTs targeted the D-form. GO's negatively charged surface provided additional interaction sites for positively charged amines in the derivatives. Similarly, studies [48] demonstrated the enantioselective absorption of phenylalanine, phenethylamine, alanine, and leucine using monolithic absorbents composed of helical polyacetylene and GO. Another group [47] designed nanoporous GO for >99% selective adsorption of L-asparagine from racemic solutions. Magnetic GO-homochiral MOF composites were also developed [45], achieving rapid and efficient enantio-separation of chiral drugs using magnetic solid-phase extraction. A strategic combination of chiral structures with other functional properties can yield innovative materials. For instance, researchers [49] integrated boronic acid groups with optically active chiral polyacetylene microspheres. These microspheres were synthesized through suspension polymerization of chiral (M1) and achiral (M2) acetylene derivatives. CD spectra confirmed the formation of helical copolymer chains, imparting enantioselective adsorption capabilities. Using L-DOPA as an adsorbate, M1 microspheres showed lower adsorption capacity but higher enantioselectivity (L/D = 1.45), whereas M2 exhibited high adsorption with negligible enantioselectivity (L/D = 1.07). Interestingly, M1-M2 copolymers offered a balance between capacity (66 mg/g for L-DOPA; 39 mg/g for D-DOPA) and enantioselectivity (L/D = 1.69). Reusability and desorption efficiency of these microspheres were also confirmed, showcasing their potential in advanced applications.

A strategic combination of chiral structures with other functional properties can yield innovative materials. For example, boronic acid groups were introduced in [49] with optically active chiral polyacetylene microspheres. These microspheres were synthesized through suspension polymerization of chiral (M1) and achiral (M2) acetylene derivatives. CD spectra confirmed the formation of helical copolymer chains, imparting enantioselective adsorption capabilities. Using L-DOPA as an adsorbate, M1 microspheres showed lower adsorption capacity but higher enantioselectivity (L/D = 1.45), whereas M2 exhibited high adsorption with negligible enantioselectivity (L/D = 1.07). Interestingly, M1-M2 copolymers offered a balance between capacity (66 mg/g for L-DOPA; 39 mg/g for D-DOPA) and enantioselectivity (L/D = 1.69). Reusability and desorption efficiency of these microspheres were also confirmed, showcasing their potential in advanced applications [49].

MOFs have emerged as a novel class of sorbents due to their extensive surface area, adjustable pore size, high sensitivity, exceptional chemical selectivity, and diverse chemical architectures [50, 51]. The pharmacokinetics underlying the separation and isolation of enantiomers is a critical factor in drug discovery and development. While numerous methods exist for separating chiral compounds, the use of MOFs has recently gained significant attention in the fields of separation and extraction science. MOFs offer unique advantages in these applications, particularly through the "host-guest" interactions they facilitate, which enhance both selectivity and resolution in the separation of chiral drugs. For example, a study [50] synthesized a composite of a chiral ionic liquid (CIL) and the MOF HKUST-1 for the enantioselective separation of three clinical drugs – atenolol, propranolol, and racecadotril – from tablet formulations. The CIL-MOF composite was developed using a tropine-based CIL containing an L-proline anion in combination with HKUST-1. The results demonstrated high efficacy (81–95%) in separating racemic mixtures of the studied drugs. Additionally, the composite exhibited excellent desorption performance and retained its reusability over five cycles. These findings highlight the promising potential of MOFs for the efficient separation of chiral compounds, including at trace levels in biological samples.

In recent years research has been actively conducted in the direction of searching for and creating new immunoaffinity adsorbents that have high affinity and selectivity for antigen-antibody interactions [52–55]. However, most of such adsorbents are unstable and the cost of their preparation is rather high. Therefore, in the present work, the authors [52] aimed to develop a new *molecular imprinted polymer* (MIP) or "synthetic antibody," which is an individual material for selective recognition of biological targets. MIP is produced by cross-linking monomers complementary in form and functional groups to the target molecule. Due to their ability to recognize a specific chemical, MIPs provide selective adsorption for the removal of various molecules such as proteins, metal ions and organic compounds. Moreover, MIPs have unique properties such as physical, chemical, mechanical and thermal stability. Their advantage is that they are more stable than antibodies and are stable at room temperature for a long time. The authors of [52] propose to develop a new type of MIP, which is a *cholesterol-based polymer nanosphere* (CPN) for adsorption of cholesterol, which is the most important physiological biomolecule. It is well known that cholesterol is a natural polycyclic lipophilic alcohol found in cell membranes of human and animals [53]. It tends to accumulate in blood vessels, leading to atherosclerosis of blood vessels and cholesterol embolism. Of course, controlling blood cholesterol levels and removing excessive amounts of cholesterol is an important task. In the current study, CPNs were prepared from cholesterol by emulsion polymerization without additional surfactant and optimized the conditions for cholesterol adsorption by varying contact time, initial concentration, ionic strength and temperature effects. Adsorption studies were carried out using solutions taken from the gastrointestinal tract. To prove the selectivity of CPNs, the authors of [54] work performed a comparison with unmodified cholesterol polymer nanospheres. *Transmission electron microscopy* (TEM) results showed that the polymerization of CPN and *non-modified polymer nanosphere* (NPN) resulted in the formation of spherical, monodisperse polymer nanospheres with particle sizes of 116 and 125 nm. The adsorption values of CPN and NPN at an initial cholesterol concentration of 100 mg/L were 11.72 and 2.89 mg/g, respectively, and the adsorption time was 30 minutes. As the analysis time increased, the number of adsorbed cholesterol molecules remained constant, indicating the formation of a monolayer on the surface of CPN. The maximum value of cholesterol adsorption was 714.17 mg/g at an initial cholesterol concentration of 2500 mg/L. The adsorbed cholesterol was 86.14% desorbed and the CPNs were successfully reused in the experiments. Thus, CPNs are promising materials for cholesterol removal.

4. Chiral substances: Metal-organic systems AgNPs-Ch and AgNPs-TChol

The interests of the authors of the chapter include the preparation of new hybrid nanomaterials of MOF type, the study of their physicochemical properties (structural, optical, dimensional, adsorption and separation), and the search for possible biomedical applications. In recent years, new supramolecular aggregates based on *silver nanoparticles* (AgNPs) and mesogenic ligands – *cholesterol* (Ch) and it is S-containing analog *thiocholesterol* (TCh) have been actively pursued [56].

Silver nanoparticles were prepared by classical borohydride reduction using the Brust-Schiffrin method (**Figure 5A**). Cholesterol, a natural biomolecule containing OH-groups and it is derivative thiocholesterol containing SH-groups were used as organic ligands-stabilizers (**Figure 5B**). The ligands have a similar structure, but the different nature of functional groups affects the binding strength to the surface silver atoms. *Transmission electron microscopy* (TEM) results showed that the average size of spherical AgNPs when stabilized by cholesterol was (5.5 ± 0.7) nm, thiocholesterol (2.7 ± 0.4) nm (**Figure 5C**). Silver is a plasmonic metal, so the appearance of *surface resonance band* (SRB) is a feature of the absorption spectra. When stabilized by cholesterol, the maximum of the SRB is observed at 430 nm and by thiocholesterol at 470 nm (**Figure 5D**). SRBs are characteristic of spherical AgNPs with a size of less than 10 nm, which agrees with the results obtained by TEM. The results of IR spectroscopy and comparison of characteristic frequencies of vibrations of functional groups of ligands with nanosystems showed that in all cases, there is a decrease in the absorption intensity of characteristic valence vibrations of functional groups of ligands, which is associated with their stabilizing ability with respect to surface silver atoms [56].

The results of quantum-chemical modeling of *(AgNPs)*$_n$*-ligand* hybrid systems were also published in our earlier works [57–59]. The calculations showed that the binding strength of the ligands to the silver cluster increases with increasing cluster size (from Ag_1 to Ag_{13}) when stabilized by thiocholesterol [59]. Summarizing the results of TEM, UV-visible spectroscopy and quantum-chemical modeling and calculations, we can draw an unambiguous conclusion: thiocholesterol is a stronger stabilizing ligand compared to cholesterol. The surface of silver atoms has a positive zeta potential (+0.3 mV, which is confirmed by X-ray photoelectron spectroscopy). The AgNPs-TCh system can be represented as a "core-shell" model by analogy with MOFs. Therefore, the formation of the

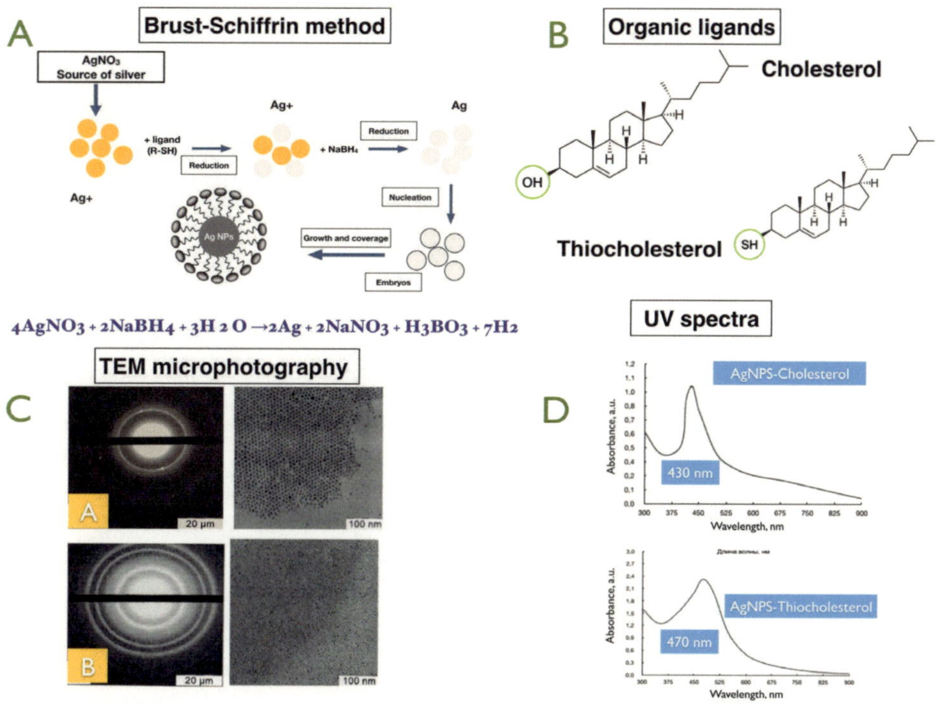

Figure 5.
A. schematic of AgNPs synthesis by Brust-Schiffrin method; B. structural formulas of cholesterol and thiocholesterol ligands; C(A). TEM-micrograph of AgNPs-Ch; C(B). TEM-micrograph of AgNPs-TCh; D. absorption spectra of AgNPs-Ch and AgNPs-TCh.

AgNPs-TCh nanosystem results in the formation of a dense "shell" with a higher dielectric constant compared to cholesterol [60, 61].

Self-ordered NPs of various metals and the strategy underlying their self-assembly provide an opportunity to analyze completely new physicochemical properties compared to their individual counterparts. The self-assembly strategy refers to a method of obtaining materials in which individual atoms/oligomers, spirals and NPs are combined into supramolecular structures due to chiral molecules. Au and Ag nanoclusters can show specific chirality arising from the nature of organic ligands or intrinsic atomic arrangement [56].

The spectra for AgNPs-TChol systems were obtained by circular dichroism (CD) spectroscopy (**Figure 6A**). To establish the regularity of the influence of the ligand content on the optical activity, samples with different molar ratios of AgNPs to TCh (1:2), (1:5) were analyzed. The incorporation of AgNPs into the crystalline matrix of TCh leads to the formation of self-ordered helical structures, induces the manifestation of the optical activity of the system and the appearance of an intense peak in the CD spectrum in the SRB of silver nanoparticles in the region of 450–470 nm, that is, the cholesteric ligand forms a spatially ordered structure that organizes AgNPs into a twisted helix. With increasing concentration of stabilized AgNPs, the optical activity of the system increases. The peaks observed at 325 and 410 nm correspond to the left-handed form of the optically active ligand thiocholesterol molecule. To prove the optical activity of the AgNPs-TCh system, the separation of enantiomers of some model analytes (**Figure 6B**) was carried out by thin-layer chromatography (TLC). For this

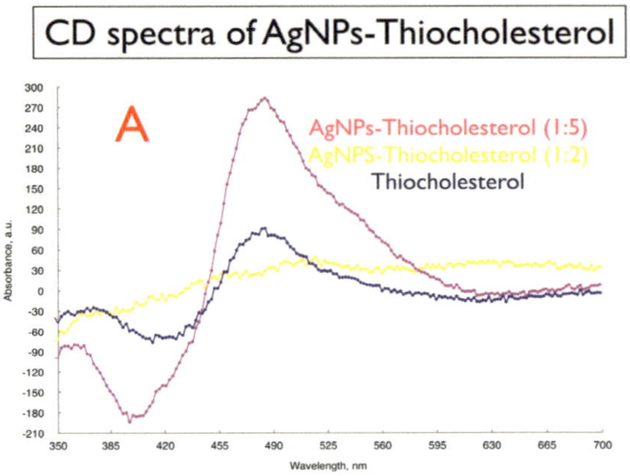

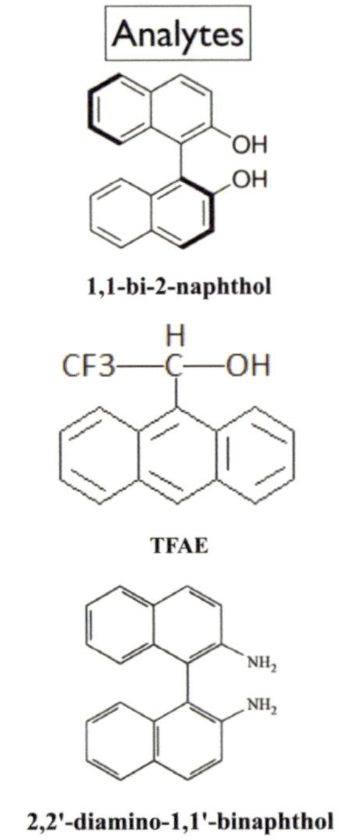

C

TLC separation results

Substance (R/S)	*Rf*	α
2,2'-diamino-1,1'-binaphthol (R)	0,57	1,56
2,2'-diamino-1,1'-binaphthol (S)	0,46	
1,1-bi-2-naphthol (R)	0,49	1,07
1,1-bi-2-naphthol (S)	0,51	
TFAE(R)	0,43	1,56
TFAE (S)	0,54	

Figure 6.
A. CD spectra of AgNPs-TCh and thiocholesterol; B. structural formulas of analytes for TLC separation; C. results of TLC separation.

purpose, the silica gel plates Silasorb-600 were modified by AgNPs-TCh solution. The separation results are presented in **Figure 6C**. To investigate the adsorption properties of the AgNPs-TCh system its impregnation from toluene solution onto silica gel carriers was carried out [60–61]. We used two types of carriers: classical silica gel (SiO_2) and aminated silica gel (SiO_2-C_3H_6-NH_2). Then we analyzed adsorption isotherms for different AgNPs-TCh systems in terms of molar ratio of metal-ligand and evaluated quantitative characteristics of the process. Optimal results were obtained for the adsorption of AgNPs-TCh in the molar ratio 1:2: the maximum amount of adsorbed AgNPs (80 mg/g) per 1 gram of silica gel was achieved at the lowest concentration (1.2 mg/ml) required for the formation of a monolayer on the adsorbent surface. On the surface of aminated silica gel, the results were more optimistic. The reason is that silver has a high affinity for NH_2 groups on the SiO_2-C_3H_6-NH_2 surface. Obviously, a higher value of the amount of adsorbed AgNPs (250 mg/g) from a solution with a concentration of 1.6 mg/ml was obtained for such sorbent.

The mechanism of surface adsorption can be described as follows: the SH-group of thiocholesterol/OH-group of cholesterol effectively interacts with the surface atoms (centers) of AgNPs during their formation and weakly shift the electron density from the cholesterol cycle. Silver has a partially positive charge, so adsorption of hybrid AgNPs is due to electrostatic interaction between positively charged surface silver atoms and silanol groups on the surface of silica gel particles [60].

Thus, it's possible to adsorb AgNPs-TCh or AgNPs-Ch system on the surface of carriers of different nature (primarily biocompatible with the human body), which

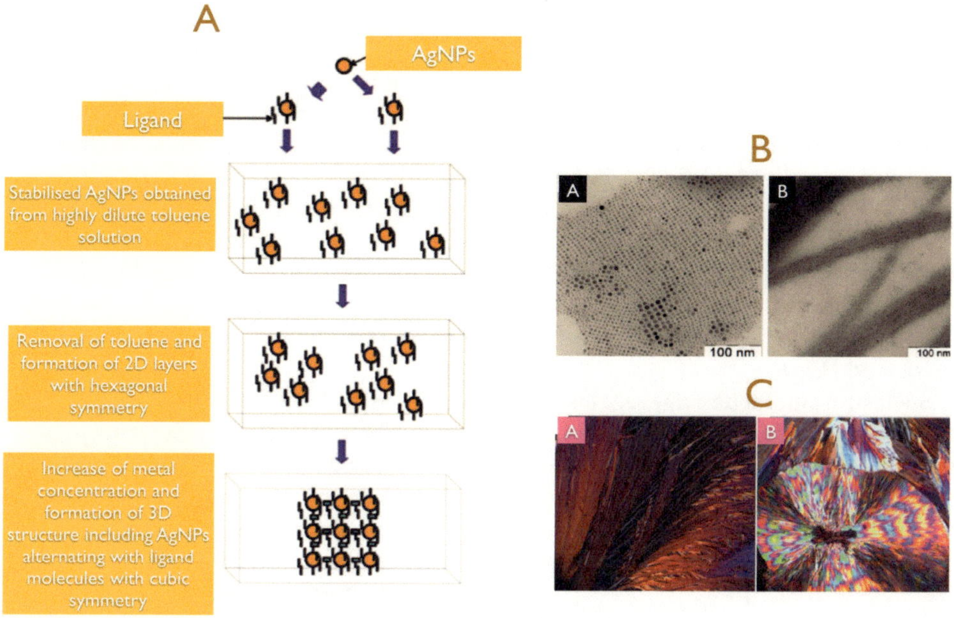

Figure 7.
A. scheme of the formation of binary nanostructure AgNPs-ligand from ternary system toluene AgNPs-ligand.; B(A), TEM micrographs of the hybrid AgNPs-TCh nanosystem 2D nanostructure with cubic symmetry, (B)-3D nanostructure obtained after solvent removal and incubation in cholesteric mesophase; C. polarization optical microscope photograph of TCh in solid phase (A) and the hybrid nanosystem AgNPs-TCh in liquid crystalline phase (B) after heating-cooling cycle, T = 25°C (x 100).

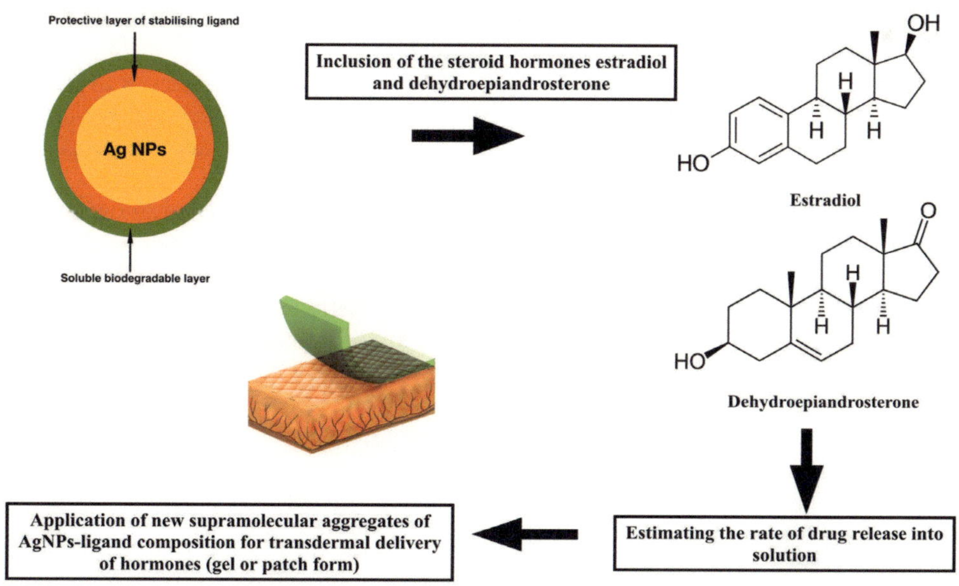

Figure 8.
Schematic showing the possibility of biomedical application of supramolecular aggregates of AgNPs-Ch and AgNPs-TCh.

is a prospect for their use as directed drug delivery systems including the transport of optically active drugs due to the optical activity proved by CD spectroscopy.

As mentioned earlier, supramolecular ordered nanostructures with pronounced optical activity are increasingly used in biomedical applications. The process of formation of ordered nanostructures can be described as follows: it starts with the formation of individual AgNPs in highly dilute organic sols, stabilized by a cholesteric ligand (**Figure 7A**). As the concentration of AgNPs increases and the organic solvent is removed, supramolecular ordered hybrid 2D and 3D nanosystems with hexagonal or cubic symmetry are formed. AgNPs are combined into ordered ensembles with chain arrangement of individual metal particles. In cholesteric mesophase, extended (L > 10 microns) linear aggregates including chain structures of AgNPs (d = 2.5 nm) are formed (**Figure 7B**) [56].

Such supramolecular aggregates can be successfully used as effective carriers of structurally similar drugs in view of their proven adsorption properties with preservation of the most important physicochemical characteristics: small size, liquid crystal phase of the ligand (**Figure 7C**), optical activity of the systems [62–65]. For example, structurally similar steroidal female and male hormones (dehydroepiandrosterone and estradiol). In this way, new forms of drug transport delivery can be created (**Figure 8**) [64].

5. Conclusions

This review fully reflects the trends and interests of modern science. The review actualizes the role of chirality in life and science. The phenomenon of chirality is considered, and classifications of modern chiral substances into organic, inorganic and hybrid materials are systematized. The authors pay much attention to organometallic nanomaterials as the most promising materials from the point of view of their practical potential including biomedical applications.

The results of the authors' work on the creation of chiral nanomaterials as promising separating, catalytic, adsorptive materials with great potential for application in biomedicine – in diagnostics, therapy and treatment of various diseases, including cancer – are presented and summarized.

The authors present the results of their research aimed at obtaining ordered supramolecular aggregates including AgNPs with ligands Ch and TCh, which exhibit optical activity. The possibility of using such materials as adsorption and separation materials with the preservation of their unique physicochemical properties is shown. The possibility of adsorption of aggregates on the surface of carriers of different nature, primarily biocompatible with the human body, is discussed, which is a prospect for their use as directed drug delivery systems, including the transport of optically active drugs due to the proven optical activity of the systems.

Acknowledgements

The authors are grateful to V.N. Morozov, Research Scientist, IBCP RAS, for the CD spectra. The authors are grateful to S.S. Abramchuk, senior researcher, for the study of samples by TEM. The authors are grateful to A.Y. Ermilov, senior researcher at Lomonosov Moscow State University, for quantum-chemical calculations.

Conflict of interest

The authors declare no conflict of interest.

Funding

The work is supported by the MSU-2024 Development Programme.

Abbreviations

NMs	nanomaterials
NPs	nanoparticles
MOMs	metal-organic materials
MOFs	metal-organic frameworks
L/D-TA L/D	tartaric acid
BPEI	branched polyethyleneimine
APTES	3-aminopropyltrietoxy silane
CNM	carbon nanomaterial
GO	graphene oxide
GO-SNTs	graphene oxide spiral silicon nanotubes
DOPA	3,4-dihydroxy-L-phenylalanine
CIL	chiral ionic liquid
MIP	molecular imprinted polymer
CPN	cholesterol-based polymer nanosphere
TEM	transmission electron microscopy
NPN	non-modified polymer nanosphere
AgNPs	silver nanoparticles
Ch	cholesterol
TCh	thiocholesterol
SRB	surface resonance band
CD	circular dichroism
TLC	thin-layer chromatography

Author details

Yana Andreevna Gromova[1] and Tatyana I. Shabatina[1,2]*

1 Department of Chemistry, Lomonosov Moscow State University, Moscow, Russia

2 Faculty of Fundamental Science, N.E. Bauman Moscow Technical State University, Moscow, Russia

*Address all correspondence to: tatyanashabtina@yandex.ru

References

[1] Yuan F, Shaobo OY, Dong Z, Junchao W, Lan L. Biological applications of chiral inorganic nanomaterials. Chirality. 2022;**34**(5):760-781. DOI: 10.1002/chir.23428

[2] Teng Y, Gu C, Chen Z, Jiang H, Xiong Y, Liu D, et al. Advances and applications of chiral resolution in pharmaceutical field. Chirality. 2022;**34**(8):1094-1119. DOI: 10.1002/chir.23453

[3] Aboul-Enein HY, Bounoua N, Rebizi M, Wagdy H. Application of nanoparticles in chiral analysis and chiral separation. Chirality. 2021;**33**(5):196-208. DOI: 10.1002/chir.23303

[4] Rossino G, Robescu MS, Licastro E, Tedesco C, Martello I, Maffei L, et al. Biocatalysis: A smart and green tool for the preparation of chiral drugs. Chirality. 2022;**34**(11):1403-1418. DOI: 10.1002/chir.23498

[5] Lin GQ, You QD, Jie-Fei C. Chiral Drugs: Chemistry and Biological Action. 2011. 10.1002/9781118075647

[6] Peng S, Zhu Y, Luo C, Zhang P, Wang F, Li R, et al. Chiral drugs: Sources, absolute configuration identification, pharmacological applications, and future research trends. LabMed Discovery. 2024;**1**(1):10008. DOI: 10.1016/j.lmd.2024.100008

[7] Garcia MC, Aloisio C, Onnainty R, Ullio-Gamboa G. Self-assembled nanomaterials. Nanobiomaterials. 2018:41-94. DOI: 10.1016/B978-0-08-100716-7.00003-9

[8] Abbas U, Li S, Liu J, Siddique X, Shi A, Hou Y, et al. Chiral metal nanostructures: Synthesis, properties and applications. Rare Metals. 2023;**42**:2489-2515. DOI: 10.1007/s12598-023-02274-4

[9] Gumus E, Bingol H, Zor E. Nanomaterials-enriched sensors for detection of chiral pharmaceuticals. Journal of Pharmaceutical and Biomedical Analysis. 2022;**221**:115031. DOI: 10.1016/j.jpba.2022.115031

[10] Cong Y, Ben Hang Y, Yu W, Shuangjiang L, W. Xinbo: Advances in membrane-based chiral separation. Coordination Chemistry Reviews. 2023;**495**:215392. DOI: 10.1016/j.ccr.2023.215392

[11] Qi Z, Song X, Ang L, Siru R. Functional materials in chiral capillary electrophoresis. Coordination Chemistry Reviews. 2021;**445**:214108. DOI: 10.1016/j.ccr.2021.214108

[12] Jiale G, Jinxing H, Juntao H, Yajiao G, Mengxue L, Hui W, et al. Recent advances in β-cyclodextrin-based materials for chiral recognition. Chemical Communications. 2023;**59**:10. DOI: 10.1039/D3CC01962D

[13] Sun M, Wang X, Guo X, Xu L, Kuang H, Xu C. Chirality at nanoscale for bioscience. Chemical Science. 2022;**13**(11):3069-3081. DOI: 10.1039/d1sc06378b

[14] Zhao R, Xiang J, Wang B, Chen L, Tan S. Recent advances in the development of noble metal NPs for cancer therapy. Bioinorganic Chemistry and Applications. 2022;**2022**:2444516. DOI: 10.1155/2022/2444516

[15] Malik P, Inwati GK, Gupta R, Mukherjee TK. Recent progress in gold and silver nanoparticle mediated drug delivery to breast cancers gold and silver nanoparticles synthesis and

applications. Micro&Nano Technologies. 2023;**9**:291-328. DOI: 10.1016/B978-0-323-99454-5.00012-3

[16] Gupta M, Seema K. Living Nanofactories: An eco-friendly approach towards medicine and environment. In: Bio-Manufactured Nanomaterials: Perspectives and Promotion. Cham, Switzerland: Springer; 2021. pp. 95-124. DOI: 10.1007/978-3-030-67223-2_6

[17] Wang J, Ni Q, Wang Y, Zhang Y, He H, Gao D, et al. Nanoscale drug delivery systems for controllable drug behaviors by multi-stage barrier penetration. Journal of Controlled Release. 2021;**10**:282-295. DOI: 10.1016/j.jconrel.2020.08.045

[18] Tian Z, Mai Y, Meng T, Gou G, Yang J. Nanocrystals for improving Oral bioavailability of drugs: Intestinal transport mechanisms and influencing factors. AAPS Pharm SciTech. 2021;**22**:179. DOI: 10.1208/s12249-021-02041-7

[19] Mok ZH. The effect of particle size on drug bioavailability in various parts of the body. Pharmaceutical Science Advances. 2023;**2**:100031. DOI: 10.1016/j.pscia.2023.100031

[20] Li JR, Sculley J, Zhou HC. Metal-organic frameworks for separations. Chemical Reviews. 2012;**112**(2):869-932. DOI: 10.1021/cr200190s

[21] Wong HL, Tao CH, Zhu N, Yam VW. Photochromic alkynes as versatile building blocks for metal alkynyl systems: Design, synthesis, and photochromic studies of diarylethene-containing platinum (II) phosphine alkynyl complexes. Inorganic Chemistry. 2011;**50**(2):471-481. DOI: 10.1021/ic101298c

[22] Wang H, Liu Y, Li J. Designer metal-organic frameworks for size-exclusion-based hydrocarbon separations: Progress and challenges. Advanced Materials. 2020;**32**(44):e2002603. DOI: 10.1002/adma.202002603

[23] Dhurjad P, Dhalaram CS, Ali N, Kumari N, Sonti R. Metal-organic frameworks in chiral separation of pharmaceuticals. Chirality. 2022;**34**(11):1419-1436. DOI: 10.1002/chir.23499

[24] Scriba G, Schiller F. Chiral separations in capillary electrophoresis. Chemistry, Molecular Sciences and Chemical Engineering. 2015:1-9. DOI: 10.1016/B978-0-12-409547-2.11499-4

[25] Xueqin H, Qiuxia C, Yiling M, Chan H, Weixia Z, Jiahao L, et al. Chiral Au nanostars for SERS sensing of enantiomers discrimination, multibacteria recognition and photothermal antibacterial application. Chemical Engineering Journal. 2023;**479**:147528. DOI: 10.1016/j.cej.2023.147528

[26] Fan L, Fang L, Yongtao Y, Yu W, Weiqun T. Strategies on nanodiagnostics and nanotherapies of the three common cancers. Nanomaterial. 2018;**8**(4):202. DOI: 10.3390/nano8040202

[27] Doughty AC. Nanomaterial applications in photothermal therapy for cancer. Materials (Basel). 2019;**12**(5):779. DOI: 10.3390/ma12050779

[28] Jeane S, Antonius W, Marvel W, Maria Y, Sandy H, Shella S, et al. Twisted-chiral mesoporous silica from coconut husk waste designed for high-performance drug uptake and sustained release. Microporous and Mesoporous Materials. 2024;**367**:112973. DOI: 10.1016/j.micromeso.2023.112973

[29] Xin W, Chang L, Na F, Sanming L, Haotian Z, Lei S, et al. Amino

functionalized chiral mesoporous silica nanoparticles for improved loading and release of poorly water-soluble drug. Asian Journal of Pharmaceutical Sciences. 2019;**14**:405-412. DOI: 10.1016/j.ajps.2018.04.002

[30] A. Cemil: A novel chiral monolithic nano-column with 50 μm i.d. for the enantio-separation of chiral drugs by nano-liquid chromatography. Green Analytical Chemistry. 2024;**8**:100094. DOI: 10.1016/j.greeac.2024.100094

[31] Xianmou G, Kaijun G, Yang H, Qi P, Yuxin W, Shun Z, et al. Biomimetic aminated mesoporous silica xerogel exhibiting chiral surface topology for delivery insoluble drug with multiple-controlled manners and improved oral adsorption. Materials and Design. 2022;**223**:111157. DOI: 10.1016/j.matdes.2022.111157

[32] Wei X, Lin W, Jiahui L, Yanbu W, Qi P, Yang H, et al. Mesoporous silica nanoparticles with chiral pattern topological structure function as “antiskid tires” on the intestinal mucosa to facilitate oral drugs delivery. Asian Journal of Pharmaceutical Sciences. 2023;**18**:100795. DOI: 10.1016/j.ajps.2023.100795

[33] Li X, Zhao L, Shao C, Li X, Sun W. Yichun Liu: Immobilization of ultrafine Ag nanoparticles on well-designed hierarchically porous silica for high-performance catalysis. Journal of Colloid and Interface Science. 2018;**530**:345-352. DOI: 10.1016/j.jcis.2018.06.045

[34] Pandey S, Ramontja J. Sodium alginate stabilized silver nanoparticles-silica nanohybrid and their antibacterial characteristics. International Journal of Biological Macromolecules. 2016;**93**:712-723. DOI: 10.1016/j.ijbiomac.2016.09.033

[35] Badri B, Yeakub Z, Terry PB. Green synthesis of gold and silver nanoparticles: Challenges and opportunities. Current Opinion in Green and Sustainable Chemistry. 2018;**12**:91-100. DOI: 10.1016/j.cogsc.2018.06.007

[36] Pang Y, Tao X, Qin Z, Jiang M, Song E, Song Y. Chiral silver nanoparticles with surface-anchored L(D)-Cys exhibit dissimilar biological characteristics in vitro but not in vivo. Toxicology Letters. 2024;**398**:28-37. DOI: 10.1016/j.toxlet.2024.06.002

[37] Baimanov D, Cai R, Chen C. Understanding the chemical nature of nanoparticle-protein interactions. Bioconjugate Chemistry. 2019;**30**(7):1923-1937. DOI: 10.1021/acs.bioconjchem.9b00348

[38] Gupta D, Roy I, Gandhi S. Metallic nanoparticles for CT-guided imaging of tumors and their therapeutic applications. OpenNano. 2023;**12**:100146. DOI: 10.1016/j.onano.2023.100146

[39] Bandi SP, Kumbhar YS, Venuganti VVK. Effect of particle size and surface charge of nanoparticles in penetration through intestinal mucus barrier. Journal of Nanoparticle Research. 2020;**22**:62. DOI: 10.1007/s11051-020-04785-y

[40] Ding S, Khan AI, Cai X, Song Y, Lyu Z, Du D, et al. Overcoming blood-brain barrier transport: Advances in nanoparticle-based drug delivery strategies. Materials Today (Kidlington). 2020;**37**:112-125. DOI: 10.1016/j.mattod.2020.02.001.550

[41] Wu L, Zhao L, Su X, Zhang P, Ling G. Repaglinide-loaded nanostructured lipid carriers with different particle sizes for improving oral absorption: Preparation, characterization, pharmacokinetics, and in situ intestinal perfusion. Drug Delivery. 2020;**27**(1):400-409. DOI: 10.1080/10717544.2019.1689313

[42] Roopali J, Amit S, Sharma PK, F. Neeraj.: Smart carbon nanotubes for drug delivery system: A comprehensive study. Journal of Drug Delivery Science and Technology. 2020;**58**:101811. DOI: 10.1016/j.jddst.2020.101811

[43] Qi W, Haitao L, Liang Z. Applications of carbon nanomaterials in chiral separation. TrAC Trends in Analytical Chemistry. 2020;**129**:115941. DOI: 10.1016/j.trac.2020.115941

[44] Gao Y, Du W, Yu A, Li X, Zhao W, Zhang S. Engineering a MOF-magnetic graphene oxide nanocomposite for enantioselective capture. Analytical Methods. 2018;**10**:5811e5816. DOI: 10.1039/c8ay01876f

[45] Ma X, Zhou X, Yu A, Zhao W, Zhang W, Zhang S, et al. Functionalized metal-organic framework nanocomposites for dispersive solid phase extraction and enantioselective capture of chiral drug intermediates. Journal of Chromatography. A. 2018;**1537**:1e9. DOI: 10.1016/j.chroma.2017.12.067

[46] Ma X, Du W, Li Y, Hua C, Yu A, Zhao W, et al. Novel chiral metal organic frameworks functionalized composites for facile preparation of optically pure propranolol hydrochlorides. Journal of Pharmaceutical and Biomedical Analysis. 2019;**172**:50e57. DOI: 10.1016/j.jpba.2019.04.034

[47] Qie F, Guo J, Tu B, Zhao X, Zhang Y, Y. Yan: b-Cyclodextrin functionalized nanoporous graphene oxides for efficient resolution of asparagine enantiomers. Chemistry, an Asian Journal. 2018;**13**:2812e2817. DOI: 10.1002/asia.201800970

[48] Li W, Wang B, Yang W, Deng J. Chiral monolithic absorbent constructed by optically active helical-substituted polyacetylene and graphene oxide: Preparation and chiral absorption capacity. Macromolecular Rapid Communications. 2015;**36**:319e326. DOI: 10.1002/marc.201400546

[49] Deng X, Liang J, Deng J. Boronic acid-containing optically active microspheres: Preparation, chiral adsorption and chirally controlled release towards drug DOPA. Chemical Engineering Journal. 2016;**306**:1162-1171. DOI: 10.1016/j.cej.2016.08.061

[50] Abbas A, Ahmad MS, Cheng YH, AlFaify S, Choi S, Irfan RM, et al. A comprehensive review on the enantiomeric separation of chiral drugs using metal-organic frameworks. Chemosphere. 2024;**364**:143083. DOI: 10.1016/j.chemosphere.2024.143083

[51] Xing K, Emily S, Benjamin SW, Ziping L, Yan L, Lujia L, et al. Yong: Challenges and opportunities for chiral covalent organic frameworks. Chemical Science. 2022;**13**:9811-9832. DOI: 10.1039/D2SC02436E

[52] Inanan T, Tüzmen N, Akgöl S, Denizli A. Selective cholesterol adsorption by molecular imprinted polymeric nanospheres and application to GIMS. International Journal of Biological Macromolecules. 2016;**92**:451-460. DOI: 10.1016/j.ijbiomac.2016.07.007

[53] Lange Y, Steck TL. Active membrane cholesterol as a physiological effector. Chemistry and Physics of Lipids. 2016;**199**:74-93. DOI: 10.1016/j.chemphyslip.2016.02.003

[54] Benesch MGK, Lewis RNAH, Ronald N. McElhaney: A calorimetric and spectroscopic comparison of the effects of cholesterol and its sulfur-containing analogs thiocholesterol and cholesterol sulfate on the thermotropic

phase behavior and organization of dipalmitoyl phosphatidylcholine bilayer membranes. Biochimica et Biophysica Acta. 2016;**1858**:168-180. DOI: 10.1016/j.bbamem.2015.11.006

[55] Singh P, Ren X, Guo T, Li W, Shakya S, He Y, et al. Biofunctionalization of cyclodextrin nanosponges using cholesterol. Carbohydrate Polymers. 2018;**190**:23-30. DOI: 10.1016/j.carbpol.2018.02.044

[56] Shabatina TI, Belyaev AA, Sergeev GB. Self-assembled nanostructures in silver-cholesterol and silver-thiocholesterol systems. BioNanoScience. 2013;**3**(3):289-294. DOI: 10.1007/s12668-013-0092-7

[57] Ermilov AY, Shabatina TI, Gromova YA. Biligand complexes of cholesterol and Thiocholesterol with silver nanoparticles: Experimental data and DFT-modeling. Moscow University Chemistry Bulletin. 2022;**77**:65-71. DOI: 10.3103/S0027131422070069

[58] Ermilov AY, Shabatina TI, Gromova YA. Interaction of copper clusters with cholesterol and Thiocholesterol: Nonempirical study. Moscow University Chemistry Bulletin. 2022;**77**:13-18. DOI: 10.55959/MSU0579-9384-2-2023-64-1-19-25

[59] Ermilov AY, Lukyanova ES, Gromova Ya A. Shabatina T.I.: Interaction of silver clusters with cholesterol ligands. Moscow University Chemistry Bulletin. 2018;**73**:251-256. DOI: 10.3103/S0027131418050048

[60] Gromova YA, Sarvin BA, Shabatina TI. Adsorption properties of hybrid metal–mesogenic silver-cholesterol and silver-thiocholesterol nanosystems. Russian Journal of Physical Chemistry A. 2019;**93**:342-347. DOI: 10.1134/S0036024419020134

[61] Shabatina TI, Gromova Ya A, Anistratova ES, Belyaev AA. New chiral metal-mesogenic nanosystems «silver-thiocholesterol» and their adsorption properties. Molecular Crystals and Liquid Crystals. 2016;**632**(1):64-69. DOI: 10.1080.15421406.2016.1185573

[62] Shabatina TI, Vernaya OI, Shimanovskiy NL, Melnikov MY. Metal and metal oxides nanoparticles and nanosystems in anticancer and antiviral theragnostic agents. Pharmaceutics. 2023;**15**:1181. DOI: 10.3390/pharmaceutics15041181

[63] Shabatina TI, Vernaya OI, Shabatin VP, Melnikov MY, Semenov AM, Lozinsky VI. Metal nanoparticle containing nanocomposites of drug substances and their potential biomedical applications. Applied Sciences. 2020;**10**(170):1-11. DOI: 10.3390/app10010170

[64] Shabatina TI, Gromova YA, Vernaya OI, Soloviev AV, Shabatin AV, Morosov YN, et al. Pharmaceutical nanoparticles formation and their physico-chemical and biomedical properties. Pharmaceuticals. 2024;**17**:587. DOI: 10.3390/ph17050587

[65] Gromova YA, Shabatina TI. Biotechnology-Biosensors, Biomaterials and Tissue Engineering-Annual Volume. London: IntechOpen; 2024. DOI: 10.5772/intechopen.115153

Chapter 3

Surface Functionalization of a Biosorbent with Surfactant and Iron Nanoparticles for Fluoride Adsorption: Mechanisms and Thermodynamic Parameters

Selene Anaid Valencia-Leal, Adriana Vázquez-Guerrero, Ruth Alfaro-Cuevas-Villanueva, Rafael Huirache-Acuña, Orlando Hernández-Cristobal and Raúl Cortés-Martínez

Abstract

The contamination of groundwater by fluoride poses a serious global health concern. This research investigates the enhancement of guava seed microparticles (GSM) for fluoride adsorption by modifying them with a cationic surfactant (HDTMA-Br) and iron nanoparticles (MNP), resulting in a functionalized biosorbent (GSM-Fe). The research evaluates the process's fluoride adsorption capacity, mechanisms, and thermodynamics. Characterization techniques confirmed successful surface modifications, including SEM, FTIR, and zeta potential measurements. Batch experiments assessed the impact of contact time, adsorbent dosage, solution pH, and temperature on fluoride removal. GSM-Fe exhibited a maximum adsorption capacity of 559.9 mg/g at 25°C, significantly higher than unmodified GSM (89.2 mg/g). The adsorption followed the Langmuir-Freundlich isotherm and pseudo-second-order kinetics. Fluoride adsorption primarily occurs through hydrogen bonding, electrostatic attraction, and interactions with functional groups such as hydroxyl and carboxyl groups on the biosorbent surface. The process was endothermic, with higher fluoride removal efficiency at elevated temperatures. Thermodynamic analysis confirmed the spontaneity of adsorption at higher temperatures. This study demonstrates that GSM-Fe is a highly efficient, cost-effective, and sustainable biosorbent for fluoride removal from water, offering a promising solution for managing fluoride contamination, particularly in developing regions.

Keywords: fluoride, adsorption mechanisms, guava seeds, surfactant, iron nanoparticles

1. Introduction

The occurrence of specific anionic species or trace elements in water for human consumption can be considered harmful when their concentrations exceed particular thresholds. Moreover, bioaccumulation raises a significant concern about its impact on human health. Fluorides are one such anionic species that can cause critical health issues when present in drinking water in deficient (<0.5 mg/L) or excessive (>1.5 mg/L) amounts [1].

Fluoride is known to be a natural contaminant of groundwater resources worldwide. It is naturally present in minerals and geological formations, including fluorite, fluorapatite, fluorspar, cryolite, topaz, muscovite, and mica [2, 3]. Fluoride ions, found in toxic organic fluorine derivatives, are anthropogenically sourced and are crucial in industries such as fertilizers, glass aluminum, electroplating, steel, semiconductors, and nuclear activities [2]. Such activities can discharge high concentrations of fluorides into the environment, increasing the risk of introducing this highly reactive element into the food chain.

Fluoride intake occurs mainly through the consumption of water, as well as through food, beverages, medicines, toothpaste, mouthwash, and cosmetics, among others. Although at low concentrations, the use and consumption of fluorides can prevent dental caries and strengthen bones [4], its ingestion can also cause the following conditions: dental fluorosis with consumption of fluoride in water between 1.5 and 2 mg/L, skeletal fluorosis, osteoporosis, brain damage, with a fluoride consumption from 3 to 6 mg/L [4, 5], adverse effects on the reproductive system in water fluoridation systems of at least 3 mg/L, neurological effects at concentrations in water greater than 3 mg/L, as well as endocrine alterations. Exceeding the recommended threshold of fluoride intake can result in detrimental health consequences for individuals across several age groups, including newborns, children, and adults. The short-term impact of fluoride on an individual is minimal, but it has the potential to accumulate in the brain over time and gradually contributes to bodily deterioration [3].

High concentrations of fluorides in groundwater have been reported in different parts of the world. Levels as high as 7.6 mg/L have been reported in India; in some regions of China, 6.20 mg/L of fluorides have been reported [6]; as well as in Sri Lanka, Spain, Brazil, Senegal, Hungary, Pakistan, Iran, the United States, Bangladesh, Argentina, and Mexico, concentrations ranging between 2.1 and 16 mg/L have been reported [7–13]. Due to the aforesaid, it is estimated that ingesting fluorides through this route represents a high risk to human health for millions of inhabitants worldwide [11, 14]. Fluoride, due to its lack of odor, taste, and transparency, is undetectable in groundwater sources, necessitating testing to determine its presence and reduce consumer risk through effective fluoride removal methods, particularly in the developing countries.

Hence, research is needed to develop novel technologies and materials to reduce fluoride levels in water for human consumption, ensuring environmental sustainability and cost-effective removal. Utilizing waste derived from lignocellulosic biomasses offers a viable alternative to commercially available inorganic-based adsorbents [15, 16]. Published research has demonstrated the efficacy of biosorbent materials, which effectively competes with traditional inorganic-based adsorbents [17, 18]. Nevertheless, there is a need to optimize the biosorbents' preparation and modification techniques to improve their adsorption capacity and selectivity and their ability to be reused across multiple cycles.

Research on inexpensive adsorbents for pollution remediation and human health care focuses on the use of biowaste, which generates approximately 100 billion metric tons annually. These materials can significantly reduce the overall cost of the adsorption process [19]. Using biosorbents in water treatment presents numerous benefits compared to traditional chemical sorbents. These materials are cheap, frequently exhibit biodegradability in natural environments, and are readily accessible and manufactured. Several biomass sources, including algae, fungi, bacteria, and several types of waste from industrial, natural, and agricultural origins, can serve as viable raw materials for obtaining biosorbents [20].

In the present context, it is known that the seeds of Guava (*Psidium guajava*) can be considered a notable byproduct generated in significant volumes through agricultural and industrial operations, particularly in Mexico [21]. The rise in guava production and consumption in recent times has resulted in the generation of significant quantities of guava seed waste, demanding proper disposal. However, this issue could have been addressed by recycling this waste as a cost-effective biomass resource [22]. The structural characteristics of guava seeds, such as their significant cellulose, lignin, and protein content, render them well-suited for their application as a biosorbent, specifically for removing anionic pollutants found in water [21].

Despite the extensive research in biosorbents' removal of pollutants, physical or chemical treatments have been proposed to enhance these biomass materials' adsorption capacities [23]. Nevertheless, it is worth noting that carbon compounds generated from biomass exhibit diverse hydrophobic surface functions, which contribute to their reduced sensitivity and dispersion in aqueous environments. Consequently, this characteristic decreases their adsorption capacity. Hence, many researchers are dedicated to exploring the modification of the biosorbents' surface with carbon compounds, particularly surfactants [20, 24–28]. Enhancing adsorbents/biosorbents by incorporating surfactants has improved their wettability and reduced surface tension. This modification enables the adsorption of various organic pollutants, including phenolics, oils, antibiotics, and azoic dyes [23, 24, 26, 29]. Hexadecyltrimethylammonium bromide (HDTMA-Br), a cationic surfactant, exhibits reduced solubility in water due to a lengthy alkyl chain in its molecular structure; such property contributes to its low toxicity to the ecosystem [30]. Hence, the utilization of this technique in the field of adsorbent surface modification has the potential to improve the efficacy of adsorbents and biosorbents toward various pollutants.

On the other hand, in recent years, magnetic nanoparticles (MNP) have gained great interest due to their potential applications in wastewater treatment and the control of environmental pollution [31], the removal of metallic and non-metallic ions from water [32], especially iron oxide nanoparticles, among which magnetite (Fe_3O_4), maghemite (γ-Fe_2O_3), and hematite (α-Fe_2O_3) stand out. Magnetic nanoparticles offer many advantages due to their unique size, easy magnetic separation and solution recovery, and large specific surface area [33]. It should be noted that, in addition to rapid separation, magnetic nanoparticles have high thermal and mechanical stabilities [34]. Another great interest in MNP is the possibility of modifying their surface to confer specific preferential adsorption properties (functionalization). However, the primary disadvantage of MNP is the lack of selectivity for adsorbate. Thus, the adsorbents impregnated with MNP can be modified on their external surface with surfactants to overcome this problem and provide adsorption systems selectivity for better performance.

Hence, the primary aim of this investigation is to assess the efficacy of surfactant-modified guava seed microparticles functionalized with iron nanoparticles in removing fluorides from aqueous solutions. Additionally, this study intends to identify the influencing variables, namely pH, adsorbent dosage, and temperature, that impact the fluoride adsorption process, characterize the biosorbent, and elucidate the main sorption mechanisms.

2. Biosorbent preparation and characterization

Guava seeds were collected, washed, dried, ground, and sieved to obtain guava seeds microparticles (GSM). A previously described technique was used to obtain iron nanoparticles (MNP) [35]. GSMs were placed in a suspension of MNP, rinsed, dried, and put in a hexadecyltrimethylammonium bromide (HDTMA-Br) solution. The obtained biosorbent was rinsed, dried, and labeled as (GSM-Fe). The process involved removing pulp residues, preparing microparticles, functionalizing GSMs with MNP and HDTMA-Br, and storing them for future fluoride removal tests.

High-vacuum scanning electron microscopy (SEM) was used to analyze the surface morphology of GSM and GSM-Fe biosorbents, identify pores and probable binding sites, and determine the elements in the biosorbent material through elemental microanalysis (EDX). Two samples of GSM and GSM-Fe were analyzed (**Figure 1**) to determine differences in their structure. In the case of GSM, it can be observed that the sample's surface presents a heterogeneous morphology, observing smooth, lamellar, and rough parts (**Figure 1a**). Furthermore, a spherical structure, which appears to be an oil gland, can be observed (**Figure 1b**). Similar structures have been

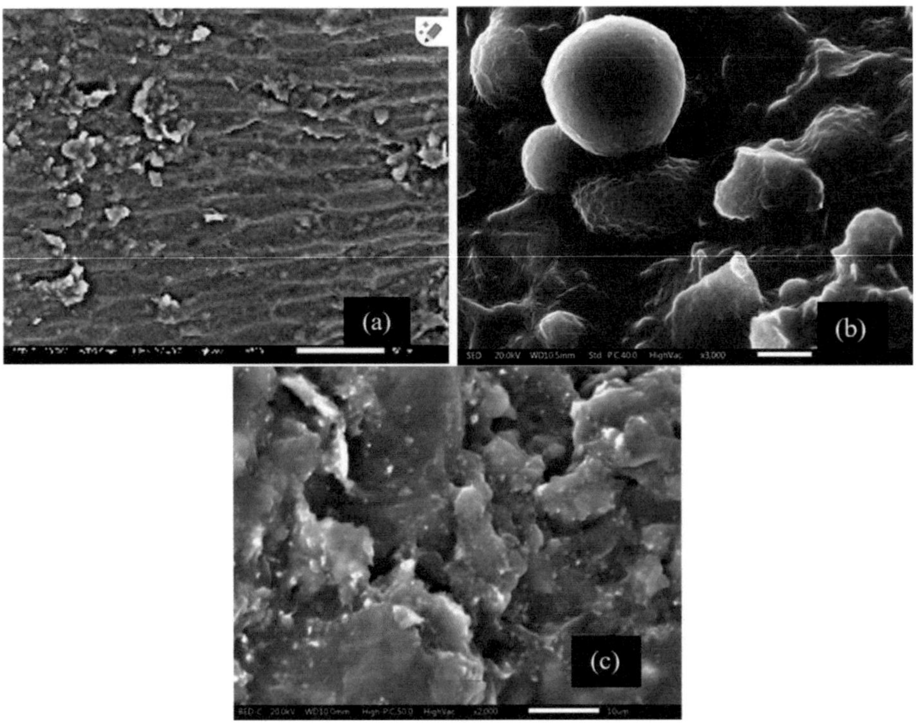

Figure 1.
SEM micrographs of (a) GSM microparticles at 500x; (b) GSM microparticles at 2000x; (c) GSM-Fe microparticles at 2000x.

reported for different seeds and tissues with high lipid and protein content [36, 37]. These compounds contain functional groups such as carboxyl, hydroxyl, esters, and amines, which help in adsorbing anions and cations in the biosorbents. These structures are advantageous for adsorbing iron nanoparticles, surfactants, and fluorides in the studied adsorption systems.

In the GSM-Fe samples, a rough, porous surface is observed (**Figure 1c**). The critical function of this type of morphology lies in its ability to enhance the surface area and, thus, the capacity for adsorbing fluoride [1]. The small white dots visible on the surface and the larger agglomerates shown in **Figure 1c** are iron nanoparticles impregnated on the surface of the GSM. This was confirmed by an EDX analysis, which revealed that iron is the primary element present. The results showed that regardless of the surface characteristics of the biosorbent, a specific amount of MNP gets attached during the modification process, leading to an approximate concentration of 1.5 g of Fe per 100 g of GSM-Fe.

A spot analysis using EDX was conducted on GSM and GSM-Fe samples to determine the presence of elements after adsorption. The analysis revealed the presence of fluorine on the surface of GSM (**Figure 2a**), a result of fluoride adsorption from an aqueous solution. The mapping of the same sample showed a slight but uniform distribution of fluorine (green dots in **Figure 2b**), highlighting the GSM's ability to remove fluoride ions. Also, an EDX analysis and surface mapping were conducted on a GSM-Fe sample after fluoride adsorption (**Figure 3a**). The analysis showed the presence of the main elements of the biosorbent, fluoride, and iron. Brighter blue spots on the surface of GSM-Fe were attributed to agglomerates of MNP adsorbed on guava seed microparticles (**Figure 3b**). The distribution and concentration of fluorine in GSM-Fe were more significant than in GSM (**Figure 3c**), suggesting an interaction between MNP and fluoride ions during adsorption. These findings suggest that fluoride can be adsorbed to iron-modified surfaces and specific unmodified structures of the biosorbent material.

FTIR analyses were performed on GSM and GSM-Fe samples before and after fluoride adsorption, as shown in **Figure 4**. The results showed that carbohydrate compounds predominate in GSM (**Figure 4a**), reflecting its complex organic structure. The material mainly contains cellulose, hemicellulose, lignin, and extractives, with the primary functional groups being carboxylic (present in hemicellulose, pectin, and lignin), amines (fats and extracts), hydroxyls (cellulose, hemicellulose, lignin, and

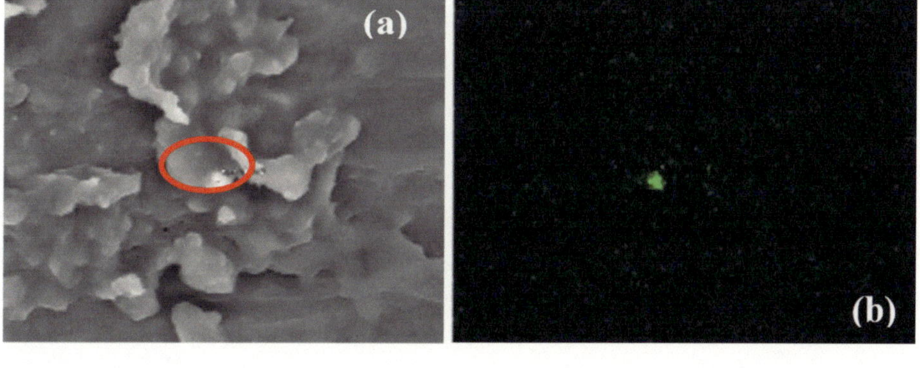

Figure 2.
SEM micrographs and EDX analysis of (a) GSM microparticles after fluoride adsorption at 1800x and (b) GSM mapping of fluorine after adsorption.

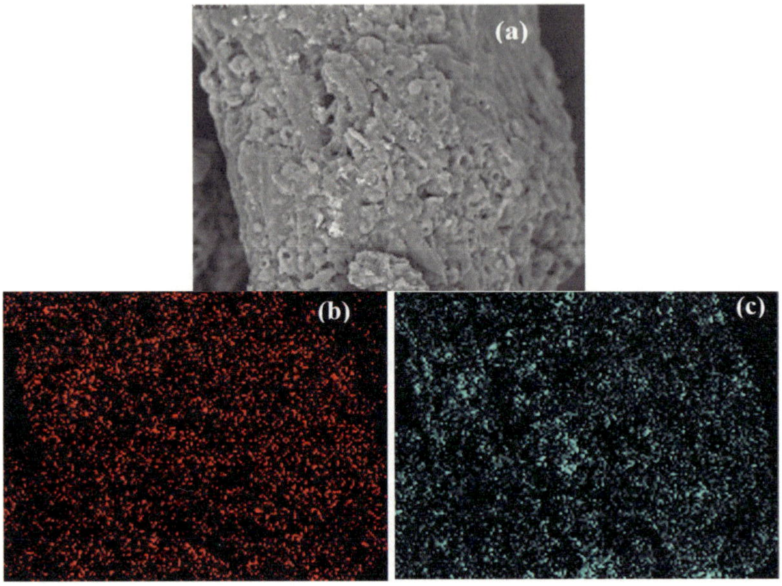

Figure 3.
SEM micrographs and EDX analysis of (a) GSM-Fe microparticles after fluoride adsorption, at 500x magnification; (b) mapping of fluorine in GSM after adsorption; and (c) Mapping of iron in GSM-Fe after adsorption.

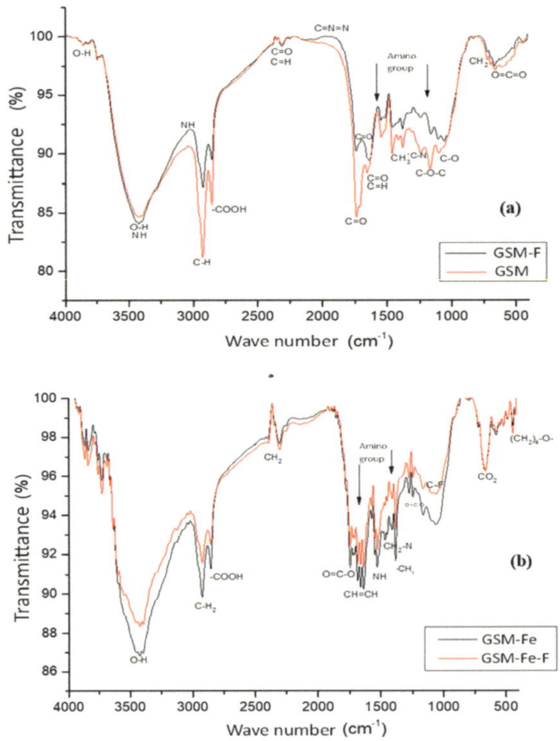

Figure 4.
FTIR spectra of (a) GSM (red line) and GSM-F after fluoride biosorption (black line); and (b) GSM-Fe (black line) and GSM-Fe-F after fluoride biosorption (red line).

pectin), and carbonyl groups (lignin, pectin) [38, 39]. The study aimed to evaluate the changes in the biosorbent's surface chemistry and understand the removal mechanisms of fluorides.

The GSM spectrum before F^- ions adsorption (red line in **Figure 4a**) shows a band at 2926.9 cm^{-1} corresponding to the C-H asymmetric stretching of aliphatic groups. At the same time, the band at 2856.2 cm^{-1} is attributed to carboxyl groups, both related to cellulose content in GSM. The broad absorption peaks at 3422 cm^{-1} are likely due to the N–H bonds in amines and amides from protein content in GSM [40–42], as well as the -OH from glucose units, primarily from cellulose and hemicellulose [1]. The C=O and C=H groups were located at the 2307 cm^{-1} small band and are related to aldehydes and ketones in GSM; no changes were noticed in these bands before or after fluoride adsorption. The band observed at 1170 cm^{-1} is assigned to the C-O-C group of the β-1,4 glycosidic links joining the sugar units of cellulose in GSM. The intensity of this peak in the GSM-F spectrum has decreased (**Figure 4a**), indicating interactions with fluorides due to surface coverage with such ions. Similar results for removing fluorides with a cellulosic biosorbent have been reported previously [43]. For GSM-F, the prominent decrease in the band observed at 2926 cm^{-1}, assigned to the stretching vibrations of the C-H group, together with an almost total reduction in the band at 1460 cm^{-1} related to the $C\text{-}H_3$ groups in aromatic rings, indicates a strong bond between fluoride ions and the cellulose chains in guava seeds. Therefore, it is likely that a C-F type bond occurs due to the interaction of fluorides with the C-H groups in cellulose [1]. The weak band at 1546 cm^{-1} has been commonly assigned to the amino (NH_2) group related to the L-tyrosine cycle [44], resulting from guava seeds' small amino acid content [45, 46]. This band also decreases in intensity after fluoride biosorption, indicating that an interaction between NH_2 and F^- ions through a hydrogen bond occurred. The functional groups observed in the GSM spectrum (**Figure 4a**) at 1676 and 1740 cm^{-1} are attributed to the carbonyl groups (C=O) associated with aldehydes and the stretching of the carboxyl group in hemicelluloses, respectively [43, 47]. The GSM-F spectrum (**Figure 4a**) shows that both peaks significantly decrease in intensity after biosorption, indicating they are involved in fluoride removal. Even the carbonyl from aldehydes shifted to 1657 cm^{-1} in GSM-F, suggesting that the halogenation of aldehydes may be an essential mechanism for removing these ions by using GSM as a biosorbent. Moreover, the peak at 2856 cm^{-1} represents the O-H stretching in carboxylic acids [48], and it also decreases after bio sorption, indicating the participation of carboxyl groups from different compounds in the GSM material in the removal of F ions, presumably by the hydrogen bonding mechanism [1].

Figure 4b presents spectra of GSM-Fe before and after fluoride adsorption. The spectrum of GSM-Fe (**Figure 4b**) indicates that this material, like GSM (**Figure 4a**), is dominated by functional groups from carbohydrate compounds, reflecting the complex organic structure. However, some bands found in GSM-Fe differ from those in GSM because the material was modified on its surface with a cationic surfactant and iron nanoparticles. Thus, bands corresponding to methylene group (CH_2) tensile vibrations are observed within the frequency range of 2301 and 2926 cm^{-1} on GSM-Fe, which are linked to the carbon chain of HDTMA-Br [49]. FTIR bands at 1712 and 1541 cm^{-1} attributed to the stretching vibrations of the acetyl groups (O-C=O) in GSM lignocellulosic biomass and to the -NH group from HDTMA-Br, respectively, support the evidence of the adsorption of surfactant with the OH groups of cellulose [50, 51]. The broad band at 3424 cm^{-1} is the characteristic of hydroxyl (-OH) groups. Still, they are also related to the presence of a layer of FeOOH on the surface of the

GSM-Fe [52], corroborating the presence of iron nanoparticles in the material. In addition, the modification with MNP results in additional peaks in GSM-Fe spectra, specifically at 1436 and 1547 cm^{-1}. These peaks have been linked to surface metal hydroxyl groups [21]. According to Ismail et al. [53], magnetite (Fe_3O_4) has two noticeable infrared absorption bands at 547 and 435 cm^{-1}. The first band corresponds to the Fe-O stretching mode of the tetrahedral and octahedral sites, whereas the band at 435 cm^{-1} corresponds to the Fe-O stretching mode of the octahedral sites. Both spectra in **Figure 4b** clearly show these bands, corroborating the MNP impregnation in the surface of GSM-Fe.

Regarding the adsorption of fluorides in GSM-Fe, the -OH groups play an essential role in removing fluorides, probably through hydrogen bonds formed between the F^- and these groups in the cellulose. In addition, it is also observed that the band corresponding to the carboxyl groups in GSM-Fe changes after the adsorption process, which indicates that the hydrogen atoms of these groups interact with the fluoride ions. Fluorides can form bonds with hydrogen atoms from carboxyl groups, as indicated by the presence of electrostatic and hydrogen bonding interactions in the adsorption process of fluoride ions. Studies have shown that fluoride binding to Tamarind seeds involves electrostatic and hydrogen bonding with phenolic/hydroxyl groups [4, 54]. The band assigned to the CH_2 groups present in the surfactant also changes after adsorption, indicating that the surfactant also adsorbs fluorides, improving the sorption capacity of GSM-Fe. Fluorides in surfactant-modified biosorbents are often linked to CH_2 groups, part of the polymer structure used to capture pollutants [55]. The CH_2 groups are frequently found in the polymer chains' backbone, which may contain quaternary amine groups. This fact enhances the sorbents' binding efficacy [55].

Zeta potential and point of zero charge (PZC) are key parameters in adsorption processes. Zeta potential influences electrostatic interactions between adsorbates and adsorbents, affecting their capacity and efficiency. PZC determines the pH of an adsorbent's surface, affecting its interaction with adsorbates and determining the nature and strength of electrostatic interactions. The zeta potential as a function of solution pH is illustrated in **Figure 5** for both GSM and GSM-Fe. It is evident that the PZC value for GSM is 3.92, while for GSM-Fe, it is 4.47. A positive surface charge develops at pH values lower than the PZC due to the presence of H^+ ions. This positive charge enhances the adsorption of anions through electrostatic attraction. In contrast, at higher pH values, the surface charge becomes negative, hindering the removal of

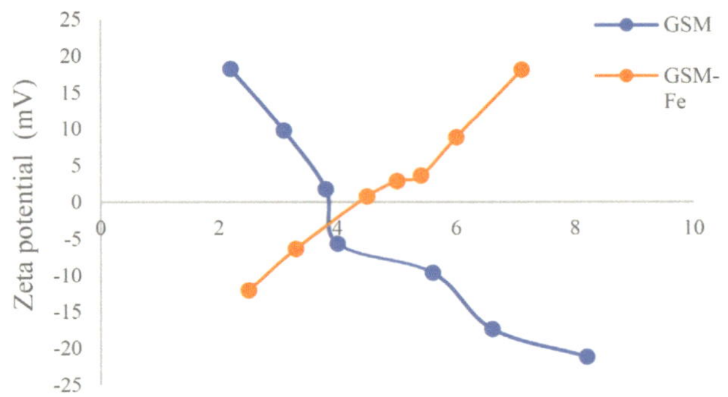

Figure 5.
Zeta potential as a function of pH for GSM and GSM-Fe.

anions. Unmodified guava seed microparticles (GSM) demonstrate this behavior, exhibiting a positive charge at lower pH levels.

GMS-Fe exhibits a negative charge at low pH values due to surface chemistry differences between GSM and GSM-Fe, as cationic surfactant modification neutralizes the surface's negative charges [56]. The parameter's positive value increases with increasing solution pH due to factors such as ionization of surface groups on particles, which affects surface charge density. Higher pH levels allow more hydroxide ions (OH^-) to adsorb on particle surfaces, increasing the negative charge [57]. This relationship is evident in several systems, including TiO_2 suspensions, where the zeta potential becomes more negative as the pH increases, indicating a higher surface charge density due to increased ionization of surface groups [58], as occurred for the GSM biosorbent. However, the presence of surfactants can modify this effect. For example, in the presence of cationic surfactants, such as benzethonium chloride (BTC), they can lead to a positive zeta potential over the entire pH range, but the effect is less pH-dependent when the surfactant concentration is high [58]; this behavior was observed for the case of the GSM-Fe biosorbent (**Figure 5**), where at higher pH values the zeta potential was more positive. The aggregation stability of suspensions also varies with pH, being higher at lower pH levels for certain surfactants, which correlates with changes in the zeta potential [58]. In addition, the pH of the solution affects the chemical speciation of metal ions and the ionization of functional groups on the adsorbent surfaces, which can influence the zeta potential. For example, functional groups are less ionized at lower pH values, leading to lower zeta potential values, while increased ionization results in higher zeta potential values [59]. Generally, the rise in zeta potential as pH increases is a complex interaction involving surface chemistry, the adsorption of ions, and specific interactions between particles and surfactants in a solution [57–59]. Those mentioned above may explain why the modification of biosorbent with HDTMA and Fe-NPs directly affected the zeta potential behavior of GSM-Fe. A similar behavior has been reported for an iron nanoparticle adsorbent modified with a similar surfactant [60].

3. Fluoride adsorption kinetics

Figure 6 shows the results obtained from the biosorption kinetics of fluorides in GSM and GSM-Fe at a temperature of 25°C and different contact times. In both biosorbents, it was observed that adsorption equilibrium is reached at approximately 120 minutes of contact. However, the fluoride removal is higher for GSM-Fe (39.8%) than for GSM under these experimental conditions. This difference in adsorption kinetics suggests that GSM-Fe could be a more effective biosorbent for fluoride removal in practical applications. In the case of GSM, the increase in adsorption as a function of time is gradual in the first minutes of contact until no significant changes in the removal are observed after 120 minutes. This behavior can be explained since, at first, the biosorbent has numerous vacant active sites, which enhances the probability of fluoride adsorption. However, the adsorption rate decreases as the sites become occupied over time. On the other hand, in the case of GSM-Fe, the increase in adsorption in the first minutes of contact is more pronounced, where at the first 20 minutes of contact, a large part of all fluoride adsorption is achieved with this biosorbent until reaching equilibrium after 120 min, indicating that the interaction

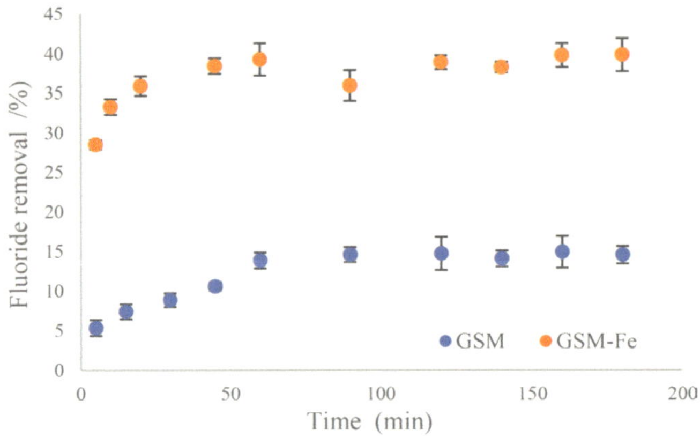

Figure 6.
Fluoride biosorption as a function of time by GSM and GSM-Fe (fluoride concentration = 10 mg/L; adsorbent dose = 30 g/L; pH = 6; temperature = 25°C).

with the iron nanoparticles and the surfactant enhances the rate at which fluorides are removed. The consistent concentration gradient in both biosorbents indicates that the initial adsorption rate difference is due to the unique surface properties of GSM-Fe, which enhances fluoride diffusion to the biosorbent's active sites.

Table 1 presents the kinetic parameters obtained through nonlinear regression to fit the experimental data into three distinct kinetic models: Lagergren, Pseudo-second-order, and Elovich [61–63]. The correlation coefficients in **Table 1** show that any empirical model can accurately describe fluoride biosorption kinetics data, with the Elovich model showing a stronger correlation for both biosorbents, indicating the presence of localized adsorption sites. According to the model, adsorption energy increases linearly as surface coverage increases, indicating a higher occupation of available adsorption sites; this implies that as more adsorbate molecules occupy available sites, the energy required for subsequent adsorption rises [64, 65]. Moreover, the parameter α is associated with the initial rate of adsorption in the system. **Table 1** data shows that this parameter is higher for GSM-Fe compared to GSM, indicating faster adsorption at the beginning of the process for GSM-Fe. On the other hand, the term $K_2.q_e$ (calculated from the pseudo-second-order equation) in adsorption kinetics is a key parameter that reflects the speed and efficiency of the adsorption process, indicating how quickly the adsorbent is capturing the adsorbate [66]. It is also directly

Biosorbent	Kinetic model		
	Elovich	Lagergren	Pseudo second order
GSM	α = 2.924 mg/g β = 0.3360 g.mg R = 0. 9622	K_L = 0.0396 min^{-1} q_e = 14.57 mg/g R = 0.9413	K_2 = 0.0034 g/mg min q_e = 16.395 mg/g $K_2.q_e$ = 0.0555 R = 0.9588
GSM-Fe	α = 7.432 mg/g β = 0.8976 g.mg R = 0.9866	K_L = 0.2504 min^{-1} q_e = 38.25 mg/g R = 0.9054	K_2 = 0.0134 g/mg h q_e = 39.41 mg/g $K_2.q_e$ = 0.5280 R = 0.9544

Table 1.
Kinetic parameters of fluoride biosorption by GSM and GSM-Fe microparticles.

related to the half-life of the adsorption process, meaning it is the inverse of the time required for half of the adsorbate to be adsorbed. This fact makes this term a helpful indicator of how fast the adsorption reaches equilibrium [66]. Based on the above, it can be established that the higher the value of this $K_2.q_e$, the shorter the time it takes for the adsorbate concentration to decrease by half; therefore, it can be concluded that this time is shorter for the case of the GSM-Fe biosorbent by one order of magnitude (**Table 1**).

4. Effect of adsorbent dosage and solution's pH

The study investigated the impact of different dosages of GSM and GSM-Fe on fluoride ion adsorption equilibrium. The experiments involved contacting different biosorbents' masses with a sodium fluoride (NaF) solution. Also, the influence of solution pH on removing fluoride ions was examined through batch-type studies using NaF solutions at varying pH levels. The initial pH values were modified by adding NaOH or HCl. Fluoride concentrations were determined in the initial solutions and supernatants. All tests were performed in triplicate.

Using the right amount of biosorbent is crucial to make the process efficient and cost-effective. Too little or too much biosorbent can result in insufficient fluoride removal, while too much can be wasteful. **Figure 7** illustrates the effect of the adsorbent dose on fluoride adsorption for both biosorbents. As expected, an increase in adsorbate removal is observed as the dosage increases in both cases due to a higher number of adsorption sites. The results showed that GSM-Fe achieved a higher percentage of fluoride removal than the other biosorbent. However, the optimal amount of GSM (30 g/L) needed for fluoride removal was slightly lower than that of GSM-Fe (40 g/L). This difference may be due to the functionalization of the biosorbent with surfactant and iron nanoparticles, which may occupy some adsorption sites in the material, requiring more adsorbent to achieve maximum removal. Additionally, if the dose is increased beyond the optimum level, the fluoride removal process no longer shows significant differences or may even start to decrease, depending on the specific biosorbent. This behavior can be attributed

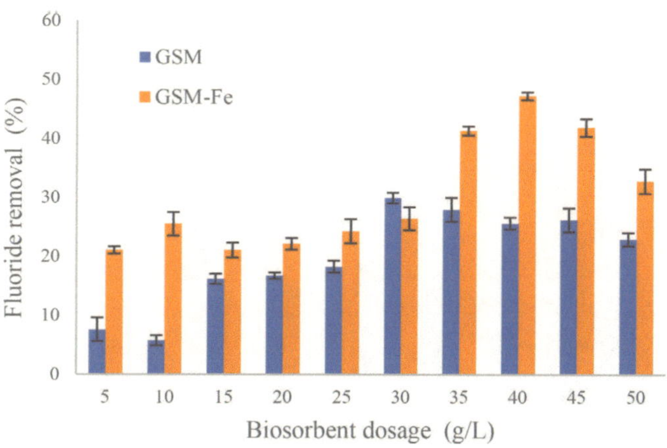

Figure 7.
Effect of adsorbent dosage on fluoride removal by GSM and GSM-Fe (fluoride concentration = 10 mg/L; pH = 6; T = 25°C).

to the higher concentration of the biosorbent particles within the batch reactor suspension mainly because of the decrease in surface area and active sites caused by particle aggregation. With an increase in the adsorbent dosage, the particles form agglomerates, reducing the available surface area for adsorption [1, 43]. These findings are significant as they clearly explain the optimal adsorbent dose required for effective fluoride removal.

On the other hand, pH plays a crucial role in adsorbate adsorption, influencing both the degree and the mechanism of adsorption; that is, pH can influence both the ionic forms of the adsorbate found in the solution and the ionized state of the functional groups responsible for its binding on the biosorbent surface. **Figure 8** shows the biosorption results of F^- ions as a function of the pH of the solution. Differences in behavior are observed between GSM and GSM-Fe. For example, for GSM, the highest removal occurred at a pH of 3, which can be attributed to the positive charge of the material at this pH value. Additionally, it can be observed that removal remains stable in a range from 5 to 7; biosorption decreases at alkaline values. The protonation equilibrium of free amino groups in aqueous solutions is favorable at a pH close to neutrality, which favors the interaction of free F^- ions in solution with this type of functional groups in GSM. In addition, FTIR findings (**Figure 4a**) support the crucial mechanism of fluoride adsorption in GSM. This equilibrium ensures the biosorbent retains its effectiveness within this pH range [67]. At alkaline pH values, there is a significant decrease in removal in GSM; this can be attributed to the material's more significant negative electrostatic charge. This greater charge causes a greater repulsion between the adsorbate and the biosorbent, which decreases diffusion from the solution to the active sites in the material. Low fluoride removal using GSM at a pH of 4 can also be observed, which aligns with the biosorbent's point of zero charge (PZC) and is the reason for this decrease; this suggests that other removal mechanisms are at play beyond the electrostatic attraction between the adsorbate and the biosorbent. The FTIR analysis findings indicated that these mechanisms could involve interactions with other functional groups in the material's structure.

Regarding the GSM-Fe material, it can be observed that it presented an effective removal of fluorides in a wide pH range (**Figure 8**). Similarly, a higher removal

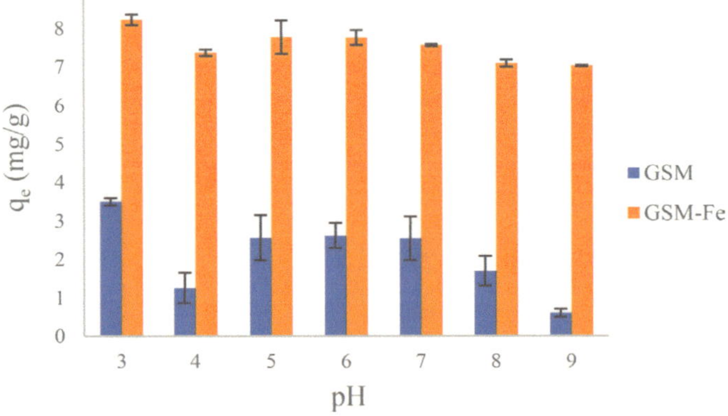

Figure 8.
Effect of initial pH on fluoride adsorption onto GSM and GSM-Fe (fluoride concentration = 10 mg/L; adsorbent dose = 30 and 40 g/L for GSM and GSM-Fe, respectively; T = 25°C).

is observed in highly acidic environments, while it decreases slightly at a pH of 4 because it is close to its PZC. The notable rise in removal under acidic conditions could be due to the creation of chemical species aside from F^- ions, such as HF and H_2F_2. The presence of HF becomes more significant as the solution's pH decreases. These species might also be adsorbed through interaction with the surfactant molecules on the outer surface of the biosorbent. Likewise, in a pH range of 5–7, no significant differences are observed in the biosorption of fluorides. At the same time, there is also a slight decrease in the removal efficiency in alkaline values (from pH 8–9). In alkaline conditions, the removal of F^- ions is much higher than in GSM; this is because, in this pH range, the electrostatic charges in GSM-Fe tend to be positive, as shown in the analysis of the zeta potential of this material (**Figure 5**). This modification introduces positive charges on the surface of the adsorbent, significantly increasing the electrostatic attraction between the adsorbent and the anionic contaminants. Moreover, modifying GSM-Fe with iron nanoparticles also enhances fluoride biosorption since, at alkaline pH, the iron oxyhydroxides on the modified biosorbent surface facilitate anion adsorption through a specific ligand exchange mechanism [68] between the F^- ions and this functional group provided by the Fe-NPs. These combined mechanisms (electrostatic attraction and ion exchange) contribute to the superior performance of GSM-Fe in removing anions in this pH range.

5. Fluoride adsorption isotherms

The study involved isotherm tests using fluoride aqueous solutions with varying concentrations, GSM and GSM-Fe samples, pH adjusted to 6, and plastic bottles for batch-type adsorption experiments. Biosorbents were contacted with NaF solutions, immersed in water baths, and agitated at different temperatures. Supernatants were filtered for fluoride analysis using the potentiometric method. **Figure 9** shows the biosorption isotherms of fluoride on GSM and GSM-Fe at different temperatures. In the case of GSM (**Figure 9a**), the adsorption capacity is the lowest at 25°C, indicating fewer adsorbate molecules in equilibrium. At 35°C, the adsorption capacity is higher but lower than that at 50°C, suggesting that increasing the temperature enhances the adsorption. At 50°C, the adsorption capacity is the highest, indicating a strong affinity between the adsorbent and adsorbate at elevated temperatures. The increase in adsorption capacity with increasing temperature suggests that the adsorption process is endothermic. Higher temperatures provide more kinetic energy to the adsorbate molecules, facilitating their movement and interaction with the adsorbent surface, and may decrease the viscosity of the solution, thereby increasing the diffusion rate of fluoride molecules through the outer boundary layer and into the internal pores of the adsorbent [69], allowing them to interact more effectively with the adsorbent surface. This increased interaction likely leads to increased adsorption capacity. Furthermore, higher temperatures may enhance the porosity or surface activity of the adsorbent through a swelling effect [70], further contributing to increased adsorption. The overall shape of the isotherms remains constant across temperatures, indicating that the fundamental adsorption mechanism does not change with temperature, although the degree of adsorption varies. The differences observed in adsorption isotherms have important implications for practical applications. This is especially true for fluorides in geothermal groundwater used for human consumption, as higher temperatures could enhance the efficiency of adsorbents in their removal. This underlines the importance of understanding and considering the

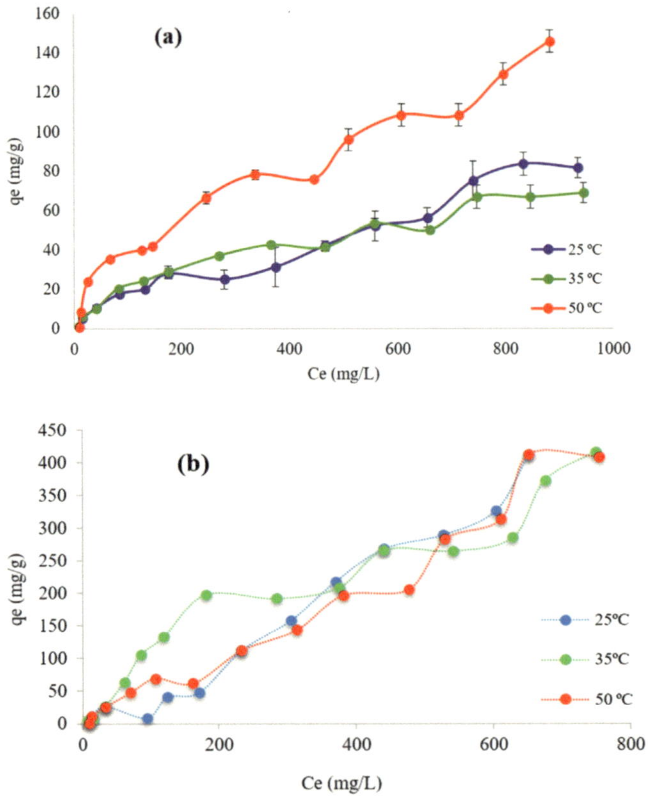

Figure 9.
Fluoride biosorption isotherms at different temperatures on (a) GSM and (b) GSM-Fe.

effects of temperature when optimizing adsorption processes in various environmental and industrial applications.

The fluoride biosorption isotherm using GSM-Fe (**Figure 9b**) shows very similar behavior to that of GSM. The adsorption capacity is higher as the solution temperature increases. Furthermore, according to Giles et al.'s classification [71], they are more similar to L-type isotherms. These isotherms are characterized by a high initial slope and a plateau, indicating strong interactions between the adsorbate and adsorbent and saturation of the adsorption sites at higher concentrations. This behavior is observed in the isotherms of both biosorbents. When comparing the isotherms of GSM and GSM-Fe, the main difference is that the fluoride adsorption capacity for GSM-Fe is significantly higher than that for GSM. This difference can be attributed to modifying guava seeds with surfactants and MNPs. Cationic surfactants, such as HDTMA-Br, enhance fluoride adsorption capacity by increasing the positive surface charge of the adsorbent. This results in higher electrostatic attraction between negatively charged fluoride ions and positively charged sites on the GMS-Fe. Cationic surfactants also introduce functional groups such as amino and imino, enhancing fluoride adsorption efficiency. The hydrophobicity of the adsorbent surface can be altered to facilitate fluoride ion adsorption by reducing competition with water molecules [72]. The modification process also increases the zeta potential, as observed for GSM-Fe (**Figure 5**), indicating higher surface charge density and improved fluoride uptake. These modifications also improve reusability, making them more effective for multiple adsorption-desorption cycles [72].

The adsorption isotherm data for each temperature were fitted to the well-known Freundlich, Langmuir, and Languir-Freundlich models by nonlinear regression to determine the equilibrium parameters of both systems under study. For each tested model, these parameters are shown in **Tables 2** and **3** for GSM and GSM-Fe, respectively. For GSM (**Table 2**), it can be observed that all models show relatively high correlation coefficients (R) (>0.9783). These high R values indicate a strong correlation between the experimental and theoretical data, enhancing the reliability of the models. There is a tendency for the Freundlich and Langmuir-Freundlich models to show better correlation, which suggests that the adsorption process involves heterogeneous surface characteristics and possibly multilayer adsorption, as well as considers the saturation of the adsorption sites. It can also be observed (**Table 2**) that it reaches a theoretical maximum adsorption capacity (Qo) of 89.25 mg/g for fluorides at 25°C, which can be considered significantly higher than other biosorbents, as observed in **Table 4**.

In the case of GSM-Fe (**Table 3**), it can also be observed that all models show relatively high R values at all temperatures tested; however, in this case, there is a higher tendency for the Langmuir-Freundlich model to describe better the adsorption equilibrium data, which suggests that the modification with MNPs and surfactant gives a more homogeneous character to the general adsorption process, without completely losing its heterogeneity. This behavior indicates that the biosorbent has a heterogeneous surface with variable site energies, which adapts to different adsorption intensities and capacities. This model is often used in practical applications where adsorbent materials are derived from natural sources or modified to improve their adsorption properties, as it can account for variability in surface characteristics and interactions between adsorbed molecules [77, 78]. Furthermore, it is also observed (**Table 3**) that the theoretical adsorption capacities (Qo) for GSM-Fe are higher than for GSM, corroborating that modification and functionalization of the biosorbent significantly improves fluoride removal from aqueous solutions at all temperatures studied. The enhanced fluoride removal, demonstrated by the higher Qo values for GSM-Fe (**Table 3**), suggests that this research could significantly impact water treatment applications. Comparison with other biosorbents (see **Table 4**) also indicates that GSM-Fe (559.93 mg/g at 25°C) is highly competitive with similar nature adsorbents and conventional adsorbents for fluoride removal under various process operating conditions.

T (°C)	Model									
	Langmuir			Freundlich			Langmuir-Freundlich			
	Q_0	b	R	K_F	n	R	K_{LF}	a_{LF}	N	R
25	81.25	0.002	0.978	0.2701	1.193	0.981	0.268	$1.52x10^{-7}$	0.839	0.981
35	97.56	0.022	0.982	1.5013	1.785	0.988	1.247	0.00053	0.607	0.988
50	233.58	0.014	0.978	2.0056	1.610	0.988	1.994	$3.1x10^{-7}$	0.622	0.988

Table 2.
Isotherm parameters for fluoride biosorption by GSM at different temperatures.

T (°C)	Langmuir			Freundlich			Langmuir-Freundlich			
	Q_0	b	R	K_F	n	R	K_{LF}	a_{LF}	N	R
25	559.93	0.00149	0.848	3.4965	1.4864	0.819	0.0000	0.0036	3.8017	0.912
35	646.59	0.00166	0.969	4.7168	1.5137	0.974	0.9583	0.00177	1.0230	0.968
50	675.01	0.00135	0.945	0.2031	0.8674	0.986	2.3439	0.00002	0.7773	0.960

Table 3.
Isotherm parameters for fluoride biosorption by GSM-Fe at different temperatures.

Biosorbent material	Adsorption capacity (Qo, mg/g)	Reference
GSM	89.25	This study
GSM-Fe	559.93	This study
Guava seeds (1 mm particle size)	15.7	[46]
Surfactant-modified banana peel	147	[20]
Tamarind seed powder	18	[54]
Iron oxyhydroxide coated with rice husk	26	[73]
Mixed-phase nano iron oxides	53.19	[60]
Zirconium-modified activated carbon fibers	28.50	[74]
Thermally treated hydrated alumina	23.7	[75]
Lanthanum-based adsorbent	245.02 (at 30°C)	[76]

Table 4.
Comparison of the theoretical maximum adsorption capacities (Qo) at 25°C for removing fluorides by various adsorbents.

6. Thermodynamic parameters

The determination of thermodynamic parameters is essential to optimize and understand adsorption processes, allowing efficiency to be improved, the best materials selected, and the design of more effective and sustainable systems to be designed. Based on fluoride adsorption data at different temperatures for GSM and GSM-Fe, the following thermodynamic parameters were determined: The change in standard Gibbs-free energy (ΔG°), standard enthalpy (ΔH°), and standard entropy (ΔS°), which were determined using the Van't Hoff equation. This method involves plotting the natural logarithm of the equilibrium constant (ln Kc) against the inverse of temperature (1/T) to derive the enthalpy change (ΔH) and entropy change (ΔS) from the slope and intercept of the graph, respectively [79]. The equilibrium constants (Kc) were calculated at different temperatures for the fluorides adsorbed by GSM and GSM-Fe. These constants are crucial for applying the Van't Hoff equation to determine thermodynamic parameters. One of our earlier works previously documented the equations related to this method [79]. Furthermore, this approach is commonly used in adsorption studies to understand the thermodynamic nature of the process.

Figure 10a and **b** show the plots of ln Kc vs. 1/T for the GSM and GS-Fe systems, respectively. These figures demonstrate a high coefficient of determination ($R^2 > 0.999$) for the data from both systems in linear regression, indicating a highly predictable pattern in the adsorption process concerning temperature changes. This predictability ensures the reliability of the method and the accurate calculation of the thermodynamic parameters from the slope and intercept of the linear graph.

Table 5 presents the thermodynamic parameters for fluoride biosorption by GSM and GSM-Fe. Notably, the values of ΔG° and ΔH are positive for both biosorbents. The Gibbs-free energy change (ΔG) is a key thermodynamic parameter that indicates the spontaneity of a process. Positive values indicate non-spontaneous processes that require an energy input. In the context of adsorption, a positive ΔG suggests that adsorption is not favorable without external energy input. At the same time, a negative ΔG indicates spontaneous processes that can occur without additional energy. However, in the case of fluoride biosorption in both systems, it is observed

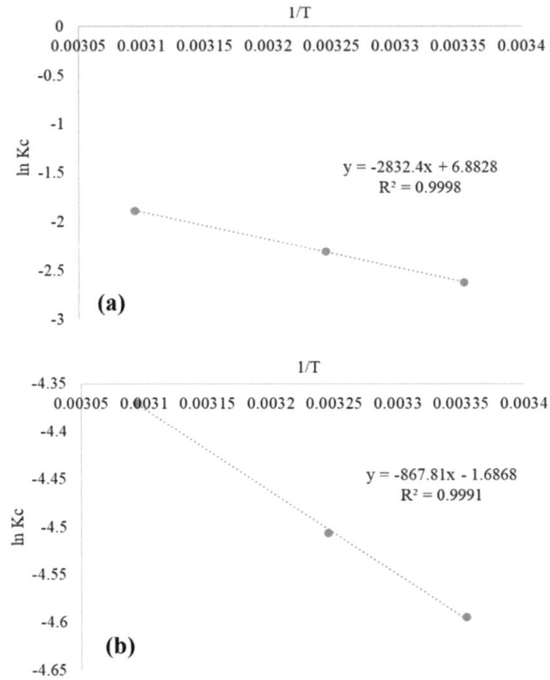

Figure 10.
Plots of ln Kc vs.1/T for fluoride biosorption on (a) GMS and (b) GSM-Fe.

that the energy required for the process to occur is sufficient at a temperature of 25°C. Furthermore, the lower ΔG value exhibited by GSM-Fe underscores that this system requires less energy for adsorption, demonstrating the potential of modifying and functionalizing guava seeds. On the other hand, the study reveals that fluoride removal by GSM and GSM-Fe is endothermic, necessitating an energy input to proceed, as the positive ΔH values (**Table 5**) indicate that the adsorption of these ions is favored at higher temperatures due to the heat absorption [79]. This endothermic nature implies that increasing the temperature can significantly enhance the adsorption capacity of biosorbents, reassuring the potential practical applications of this research.

In the case of the change in standard entropy (ΔS°), it can be observed that for GSM, the value is positive, while for GSM-Fe, it is negative (**Table 5**). A positive ΔS value indicates an increase in randomness or disorder at the solid-solution interface due to a more dispersed energy distribution between molecules

T (°C)	GSM			GSM-Fe		
	ΔG° (J/mol)	ΔH° (J/mol)	ΔS° (J/mol K)	ΔG° (J/mol)	ΔH° (J/mol)	ΔS° (J/mol K)
25	21545.74	23548.5	57.22	11396.24	7214.9	−14.02
35	22117.98			11536.48		
50	20687.39			11746.84		

Table 5.
Thermodynamic parameters for fluoride biosorption by GSM and GSM-Fe.

during adsorption [80], suggesting that the adsorption process is favorable and potentially reversible. In contrast, a negative ΔS value upon adsorption indicates that adsorbed molecules such as fluoride ions become more ordered upon the biosorbent surface, indicating that the ions are more structured or aligned on the adsorbent surface. It has been established earlier that modifying and functionalizing guava seeds can enhance the uniformity of the biosorption process. This is because a reduction in entropy usually occurs when the adsorbate forms a more organized structure on the adsorbent, possibly due to specific interactions or bonds. This highlights the effectiveness of using GSM-Fe for fluoride removal, as homogeneous adsorption processes offer advantages in water treatment. They ensure consistent performance by providing uniform adsorption sites, leading to predictable and consistent adsorption capacity.

7. Conclusions

Modifying guava seed microparticles with iron nanoparticles and cationic surfactant (GSM-Fe) significantly improved the fluoride adsorption capacity compared to unmodified microparticles (GSM). This improvement is attributed to the increase in active sites on the surface and the presence of iron nanoparticles, which creates more active sites for fluoride binding. The main mechanisms of fluoride adsorption were hydrogen bonding, electrostatic attraction, and interactions with functional groups such as hydroxyl and carboxyl groups present on the biosorbent's surface. Additionally, the presence of iron nanoparticles contributed to enhanced adsorption through ligand exchange and increased active sites for fluoride binding. Furthermore, the study demonstrated that GSM and GSM-Fe exhibited optimal fluoride removal within a pH range of 3–7. GSM-Fe showed a wider range of effective pH due to the presence of iron nanoparticles. On the other hand, biosorption kinetics for both materials was relatively fast and fit well with the Elovich and pseudo-second-order models, with GSM-Fe showing faster adsorption rates and higher overall fluoride removal. The Langmuir-Freundlich isotherm model provided the best fit for GSM-Fe, indicating that the biosorbent modification created a more homogeneous surface for fluoride adsorption, further improving efficiency. Thermodynamic analysis reveals that fluoride adsorption on GSM and GSM-Fe is endothermic, with higher temperatures favoring the process. GSM-Fe shows spontaneous adsorption, while GSM requires energy for efficient fluoride removal. The negative entropy change for GSM-Fe indicates increased fluoride ion orderliness, confirming enhanced fluoride binding capacity due to surface modification. The study reveals that guava seed biosorbents, specifically GSM-Fe, can effectively remove fluoride from aqueous solutions; this makes GSM-Fe a highly competitive biosorbent with scalability potential for continuous water treatment applications, particularly in treating fluoride-contaminated groundwater. The study supports the hypothesis that surface modification of guava seed biosorbents improves their effectiveness.

Acknowledgements

The authors acknowledge technical support from D.M. Bocanegra-González and R. Cisneros-Magaña and financial support from Coordinación de la Investigación Científica-UMSNH (grant: CIC-UMSNH-2024).

Conflict of interest

The authors declare no conflict of interest.

Abbreviations

GSM	guava seed microparticles
GSM-Fe	guava seed microparticles functionalized with surfactant and iron nanoparticles
HDTMA-Br	hexadecyltrimethylammonium bromide (surfactant)
SEM	scanning electron microscopy
FTIR	Fourier transform infrared spectroscopy
MNP	magnetic iron nanoparticles
EDX	energy-dispersive X-ray spectroscopy
PZC	point of zero charge
NaF	sodium fluoride
HCl	hydrochloric acid
NaOH	sodium hydroxide
Qe	adsorption capacity at equilibrium
R	correlation coefficient
R^2	determination coefficient
K_L	Lagergren rate constant
K_2	second-order kinetic constant
α	Elovich equation constant
β	Elovich equation constant
Qo	maximum adsorption capacity
b	Langmuir constant
K_F	Freundlich constant
K_{LF}	Langmuir-Freundlich constant
a_{LF}	Langmuir-Freundlich constant
Kc	equilibrium constant
T	temperature
$\Delta G°$	Gibbs-free energy change
$\Delta H°$	enthalpy change
$\Delta S°$	entropy change

Author details

Selene Anaid Valencia-Leal[1], Adriana Vázquez-Guerrero[2], Ruth Alfaro-Cuevas-Villanueva[1], Rafael Huirache-Acuña[3], Orlando Hernández-Cristobal[4] and Raúl Cortés-Martínez[1]*

1 Earth Sciences Research Institute, Michoacan University of San Nicolas of Hidalgo, Morelia, Michoacan, Mexico

2 Chemical Engineering Faculty, Juarez Autonomous University of Tabasco, Campus Cunduacan, Tabasco, Mexico

3 Chemical Engineering Faculty, Michoacan University of San Nicolas of Hidalgo, Morelia, Mexico

4 National School of Higher Studies (ENES), Morelia Unit, UNAM, Morelia, Michoacan, Mexico

*Address all correspondence to: raul.cortes@umich.mx

References

[1] Vázquez-Guerrero A, Alfaro-Cuevas-Villanueva R, Rutiaga-Quiñones JG, Cortés-Martínez R. Fluoride removal by aluminum-modified pine sawdust: Effect of competitive ions. Ecological Engineering. 2016;**94**:365-379. DOI: 10.1016/j.ecoleng.2016.05.070

[2] Ghaderpoori M, Paydar M, Zarei A, Alidadi H, Najafpoor AA, Gohary AH, et al. Health risk assessment of fluoride in the water distribution network of Mashhad, Iran. Human and Ecological Risk Assessment: An International Journal. 2019;**25**(4):851-862. DOI: 10.1080/10807039.2018.1453297

[3] Ahmad S, Singh R, Arfin T, Neeti K. Fluoride contamination, consequences and removal techniques in water: A review. Environmental Science: Advances. 2022;**1**(5):620-661. DOI: 10.1039/D1VA00039J

[4] Wei Y, Wang L, Li H, Yan W, Feng J. Synergistic fluoride adsorption by composite adsorbents synthesized from different types of materials—A review. Frontiers in Chemistry. 2022;**10**:900660. Available from: https://www.frontiersin.org/articles/10.3389/fchem.2022.900660 [Accessed: October 22, 2023]

[5] Jiménez-Becerril J, Solache-Ríos M, García-Sosa I. Fluoride removal from aqueous solutions by Boehmite. Water, Air, and Soil Pollution. 2012;**223**(3):1073-1078. DOI: 10.1007/s11270-011-0925-3

[6] Guo Q, Wang Y, Ma T, Ma R. Geochemical processes controlling the elevated fluoride concentrations in groundwaters of the Taiyuan Basin, northern China. Journal of Geochemical Exploration. 2007;**93**(1):1-12. DOI: 10.1016/j.gexplo.2006.07.001

[7] Carrillo-Rivera JJ, Cardona A, Edmunds WM. Use of abstraction regime and knowledge of hydrogeological conditions to control high-fluoride concentration in abstracted groundwater: San Luis Potosí basin, Mexico. Journal of Hydrology. 2002;**261**(1):24-47. DOI: 10.1016/S0022-1694(01)00566-2

[8] Armienta MA, Segovia N. Arsenic and fluoride in the groundwater of Mexico. Environmental Geochemistry and Health. 2008;**30**(4):345-353. DOI: 10.1007/s10653-008-9167-8

[9] Reddy DV, Nagabhushanam P, Sukhija BS, Reddy AGS, Smedley PL. Fluoride dynamics in the granitic aquifer of the Wailapally watershed, Nalgonda District, district. Chemical Geology. 2010;**269**(3):278-289. DOI: 10.1016/j.chemgeo.2009.10.003

[10] Keramati H, Miri A, Baghaei M, Rahimizadeh A, Ghorbani R, Fakhri Y, et al. Fluoride in Iranian drinking water resources: A systematic review, meta-analysis and non-carcinogenic risk assessment. Biological Trace Element Research. 2019;**188**(2):261-273. DOI: 10.1007/s12011-018-1418-7

[11] Kumar PS, Suganya S, Srinivas S, Priyadharshini S, Karthika M, Karishma Sri R, et al. A review. Environmental Chemistry Letters. 2019;**17**(4):1707-1726. DOI: 10.1007/s10311-019-00906-9

[12] Chicas SD, Omine K, Prabhakaran M, Sunitha TG, Sivasankar V. High fluoride in groundwater and associated non-carcinogenic risks at Tiruvannamalai region in Tamil Nadu, India. Ecotoxicology and Environmental Safety. 2022;**233**:113335. DOI: 10.1016/j.ecoenv.2022.113335

[13] Huang L, Sun Z, Zhou A, Bi J, Liu Y. Source and enrichment mechanism of fluoride in groundwater of the Hotan oasis within the Tarim Basin, northwestern China. Environmental Pollution. 2022;**300**:118962. DOI: 10.1016/j.envpol.2022.118962

[14] Podgorski J, Berg M. Global analysis and prediction of fluoride in groundwater. Nature Communications. 2022;**13**(1):4232. DOI: 10.1038/s41467-022-31940-x

[15] Robledo-Peralta A, Torres-Castañón LA, Rodríguez-Beltrán RI, Reynoso-Cuevas L. Lignocellulosic biomass as sorbent for fluoride removal in drinking water. Polymers. 2022;**14**(23):5219. DOI: 10.3390/polym14235219

[16] Huang L, Luo Z, Huang X, Wang Y, Yan J, Liu W, et al. Applications of biomass-based materials to remove fluoride from wastewater: A review. Chemosphere. 2022;**301**:134679. DOI: 10.1016/j.chemosphere.2022.134679

[17] Aryal R, Poudel B, Gautam S, Paudyal H, Ghimire K. Removal of fluoride from aqueous solution using biomass-based adsorbents: A review. Journal of Nepal Chemical Society. 2019;**40**:44-51. DOI: 10.3126/jncs.v40i0.27281

[18] Tran HN, Nguyen HC, Woo SH, Nguyen TV, Vigneswaran S, Hosseini-Bandegharaei A, et al. Removal of various contaminants from water by renewable lignocellulose-derived biosorbents: A comprehensive and critical review. Critical Reviews in Environmental Science and Technology. 2019;**49**(23):2155-2219. DOI: 10.1080/10643389.2019.1607442

[19] Saravanan A, Karishma S, Kumar PS, Thamarai P, Yaashikaa PR. Recent insights into mechanism of modified bio-adsorbents for the remediation of environmental pollutants. Environmental Pollution. 2023;**339**:122720. DOI: 10.1016/j.envpol.2023.122720

[20] Barboza JAT, Penido ES, Ferreira GMD. Production of surfactant-modified banana peel biosorbents applied to treatment and decolorization of effluents. Colloids and Surfaces A: Physicochemical and Engineering Aspects. 2024;**680**:132650. DOI: 10.1016/j.colsurfa.2023.132650

[21] Ramos-Vargas S, Alfaro-Cuevas-Villanueva R, Huirache-Acuña R, Cortés-Martínez R. Removal of fluoride and arsenate from aqueous solutions by aluminum-modified guava seeds. Applied Sciences. 2018;**8**(10):1807. DOI: 10.3390/app8101807

[22] Cai H, Chen G, Peng C, Zhang Z, Dong Y, Shang G, et al. Removal of fluoride from drinking water using tea waste loaded with Al/Fe oxides: A novel, safe and efficient biosorbent. Applied Surface Science. 2015;**328**:34-44. DOI: 10.1016/j.apsusc.2014.11.164

[23] Karaman C, Karaman O, Show P-L, Karimi-Maleh H, Zare N. Congo red dye removal from aqueous environment by cationic surfactant modified-biomass derived carbon: Equilibrium, kinetic, and thermodynamic modeling, and forecasting via artificial neural network approach. Chemosphere. 2022;**290**:133346. DOI: 10.1016/j.chemosphere.2021.133346

[24] Ansari R, Seyghali B, Mohammad-khah A, Zanjanchi MA. Application of Nano surfactant modified biosorbent as an efficient adsorbent for dye removal. Separation Science and Technology. 2012;**47**(12):1802-1812. DOI: 10.1080/01496395.2012.658485

[25] Chen G-Q, Zhang W-J, Zeng G-M, Huang J-H, Wang L, Shen G-L. Surface-modified Phanerochaete chrysosporium as a biosorbent for Cr(VI)-contaminated wastewater. Journal of Hazardous Materials. 2011;**186**(2):2138-2143. DOI: 10.1016/j.jhazmat.2010.12.123

[26] Guler UA, Solmaz B. Biosorption of tetracycline and cephalexin onto surfactant-modified waste biomass using response surface methodology and Ecotoxicological assessment: Phytotoxicity and biotoxicity studies. Water, Air, and Soil Pollution. 2022;**233**(4):117. DOI: 10.1007/s11270-022-05590-0

[27] Namasivayam C, Sureshkumar MV. Removal of chromium(VI) from water and wastewater using surfactant modified coconut coir pith as a biosorbent. Bioresource Technology. 2008;**99**(7):2218-2225. DOI: 10.1016/j.biortech.2007.05.023

[28] Ranasinghe SH, Navaratne AN, Priyantha N. Enhancement of adsorption characteristics of Cr(III) and Ni(II) by surface modification of jackfruit peel biosorbent. Journal of Environmental Chemical Engineering. 2018;**6**(5):5670-5682. DOI: 10.1016/j.jece.2018.08.058

[29] Hamidon TS, Adnan R, Haafiz MKM, Hussin MH. Cationic surfactant-modified cellulose Nanocrystal/alginate hydrogel beads for enhanced adsorptive removal of 4-Chlorophenol from wastewater. Journal of Polymers and the Environment. 2022;**30**(12):5024-5048. DOI: 10.1007/s10924-022-02573-4

[30] García MT, Ribosa I, Guindulain T, Sánchez-Leal J, Vives-Rego J. Fate and effect of monoalkyl quaternary ammonium surfactants in the aquatic environment. Environmental Pollution. 2001;**111**(1):169-175. DOI: 10.1016/S0269-7491(99)00322-X

[31] Wang P, Ma Q, Hu D, Wang L. Removal of reactive blue 21 onto magnetic chitosan microparticles functionalized with polyamidoamine dendrimers. Reactive and Functional Polymers. 2015;**91-92**:43-50. DOI: 10.1016/j.reactfunctpolym.2015.04.007

[32] Jha DK, Shameem M, Patel AB, Kostka A, Schneider P, Erbe A, et al. Simple synthesis of superparamagnetic magnetite nanoparticles as highly efficient contrast agent. Materials Letters. 2013;**95**:186-189. DOI: 10.1016/j.matlet.2012.12.096

[33] Davudabadi Farahani M, Shemirani F. Supported hydrophobic ionic liquid on magnetic nanoparticles as a new sorbent for separation and preconcentration of lead and cadmium in milk and water samples. Microchimica Acta. 2012;**179**(3):219-226. DOI: 10.1007/s00604-012-0885-3

[34] Li G, Zhou Z, Li Y, Huang K, Zhong M. Surface functionalization of chitosan-coated magnetic nanoparticles for covalent immobilization of yeast alcohol dehydrogenase from Saccharomyces cerevisiae. Journal of Magnetism and Magnetic Materials. 2010;**322**(24):3862-3868. DOI: 10.1016/j.jmmm.2010.08.008

[35] Safarik I, Safarikova M. One-step magnetic modification of non-magnetic solid materials. International Journal of Materials Research. 2013;**105**(1):104-107. DOI: 10.3139/146.111009

[36] Cardinali FJ, Thevenon MA. Estudio morfoanatómico de la semilla y de las reservas proteicas y lipídicas en tejidos cotiledonales de Cuphea Glutinosa (Lythraceae). 2010. Boletin de la

Sociedad Argentina de Botánica. 2010;**45**(1-2):47-55

[37] García-Fajardo JA, M Del R R-G, Mora-Galindo J. Estructura de la semilla de aguacate y cuantificación de la grasa extraída por diferentes técnicas. Revista Chapingo serie horticultura. 1999;**5**:123-128

[38] Wahab MA, Jellali S, Jedidi N. Ammonium biosorption onto sawdust: FTIR analysis, kinetics and adsorption isotherms modeling. Bioresource Technology. 2010;**101**(14):5070-5075. DOI: 10.1016/j.biortech.2010.01.121

[39] Tandy S, Healey JR, Nason MA, Williamson JC, Jones DL, Thain SC. FT-IR as an alternative method for measuring chemical properties during composting. Bioresource Technology. 2010;**101**(14):5431-5436. DOI: 10.1016/j.biortech.2010.02.033

[40] Farinella NV, Matos GD, Arruda MAZ. Grape bagasse as a potential biosorbent of metals in effluent treatments. Bioresource Technology. 2007;**98**(10):1940-1946. DOI: 10.1016/j.biortech.2006.07.043

[41] Sulejmanović J, Gostevčić A, Karadža A, Zahirović A, Topčagić A, Ostojić J, et al. Synergetic removal of hazardous pollutants from aqueous environment using lignocellulosic biosorbents. Journal of Molecular Liquids. 2023;**389**:122860. DOI: 10.1016/j.molliq.2023.122860

[42] Hiew BYZ, Lee LY, Lee XJ, Thangalazhy-Gopakumar S, Gan S. Utilisation of environmentally friendly okara-based biosorbent for cadmium(II) removal. Environmental Science and Pollution Research. 2021;**28**(30):40608-40622. DOI: 10.1007/s11356-020-09594-3

[43] Egor M, Kumar AA, Ahuja T, Mukherjee S, Chakraborty A, Sudhakar C, et al. Cellulosic ternary Nanocomposite for affordable and sustainable fluoride removal. ACS Sustainable Chemistry and Engineering. 2021;**9**(38):12788-12799. DOI: 10.1021/acssuschemeng.1c03272

[44] Elwardany RE, Shokry H, Mustafa AA, Ali AE. Influence of the prepared activated carbon on cellulose acetate for malachite green dye removal from aqueous solution. Macromolecular Research. 2023;**31**(11):1043-1060. DOI: 10.1007/s13233-023-00187-w

[45] Elizalde-González MP, Hernández-Montoya V. Guava seed as an adsorbent and as a precursor of carbon for the adsorption of acid dyes. Bioresource Technology. 2009;**100**(7):2111-2117. DOI: 10.1016/j.biortech.2008.10.056

[46] Sánchez-Sánchez HA, Cortés-Martínez R, Alfaro-Cuevas R. Fluoride removal from aqueous solutions by mechanically modified guava seeds. International Journal of Sciences. 2013;**11**(1):159-172

[47] Pejić B, Vukčević M, Lazić B, Janjić S, Kostić M. The role of cellulosic and noncellulosic functional groups in the biosorption of lead ions by waste flax fibers. Journal of Natural Fibers. 2023;**20**(1):2140325. DOI: 10.1080/15440478.2022.2140325

[48] Pradhan D, Sukla LB, Mishra BB, Devi N. Biosorption for removal of hexavalent chromium using microalgae *Scenedesmus* sp. Journal of Cleaner Production. 2019;**209**:617-629. DOI: 10.1016/j.jclepro.2018.10.288

[49] Nodehi R, Shayesteh H, Kelishami AR. Enhanced adsorption of Congo red using cationic surfactant functionalized zeolite particles.

Microchemical Journal. 2020;**153**:104281. DOI: 10.1016/j.microc.2019.104281

[50] Kurniawan A, Sutiono H, Indraswati N, Ismadji S. Removal of basic dyes in binary system by adsorption using rarasaponin–bentonite: Revisited of extended Langmuir model. Chemical Engineering Journal. 2012;**189-190**:264-274. DOI: 10.1016/j.cej.2012.02.070

[51] Boufi S, Alila S. Modified cellulose fibres as a biosorbent for the organic pollutants. In: Biopolymers. United Kingdom: John Wiley and Sons, Ltd; 2011. pp. 483-524. DOI: 10.1002/9781118164792.ch17. Available from: https://onlinelibrary.wiley.com/doi/abs/10.1002/9781118164792.ch17

[52] Ahmad M, Akanji MA, Usman ARA, Al-Farraj ASF, Tsang YF, Al-Wabel MI. Turning date palm waste into carbon nanodots and nano zerovalent iron composites for excellent removal of methylthioninium chloride from water. Scientific Reports. 2020;**10**(1):16125. DOI: 10.1038/s41598-020-73097-x

[53] Ismail AM, Tiama TM, Farghaly A, Elhaes H, Ibrahim MA. Assessment of the functionalization of chitosan/iron oxide nanoparticles. Biointerface Research in Applied Chemistry. 2023;**13**(6):582. DOI: 10.33263/BRIAC136.582

[54] Murugan M, Subramanian E. Studies on defluoridation of water by tamarind seed, an unconventional biosorbent. Journal of Water and Health. 2006;**4**(4):453-461. DOI: 10.2166/wh.2006.0029

[55] Tan X, Sawczyk M, Chang Y, Wang Y, Usman A, Fu C, et al. Revealing the molecular-level interactions between cationic fluorinated polymer sorbents and the major PFAS pollutant PFOA. Macromolecules. 2022;**55**(3):1077-1087. DOI: 10.1021/acs.macromol.1c02435

[56] Dayananda D, Sarva VR, Prasad SV, Arunachalam J, Ghosh NN. Preparation of CaO loaded mesoporous Al2O3: Efficient adsorbent for fluoride removal from water. Chemical Engineering Journal. 2014;**248**:430-439. DOI: 10.1016/j.cej.2014.03.064

[57] Vinogradov J, Jackson MD, Chamerois M. Zeta potential in sandpacks: Effect of temperature, electrolyte pH, ionic strength and divalent cations. Colloids and Surfaces A: Physicochemical and Engineering Aspects. 2018;**553**:259-271. DOI: 10.1016/j.colsurfa.2018.05.048

[58] Petryshyn RS, Yaremko ZM, Soltys MN. Effects of surfactants and pH of medium on zeta potential and aggregation stability of titanium dioxide suspensions. Colloid Journal. 2010;**72**(4):517-522. DOI: 10.1134/S1061933X10040125

[59] Phuengphai P, Singjanusong T, Kheangkhun N, Wattanakornsiri A. Removal of copper(II) from aqueous solution using chemically modified fruit peels as efficient low-cost biosorbents. Water Science and Engineering. 2021;**14**(4):286-294. DOI: 10.1016/j.wse.2021.08.003

[60] Mohapatra M, Rout K, Singh P, Anand S, Layek S, Verma HC, et al. Fluoride adsorption studies on mixed-phase nano iron oxides prepared by surfactant mediation-precipitation technique. Journal of Hazardous Materials. 2011;**186**(2):1751-1757. DOI: 10.1016/j.jhazmat.2010.12.076

[61] Ho YS, McKay G, Wase DAJ, Forster CF. Study of the sorption of divalent metal ions on to peat. Adsorption Science and Technology.

2000;**18**(7):639-650. DOI: 10.1260/0263617001493693

[62] Low MJD. Kinetics of chemisorption of gases on solids. Chemical Reviews. 1960;**60**(3):267-312. DOI: 10.1021/cr60205a003

[63] Lagergren S. Zur theorie der sogenannten adsorption geloster stoffe. Kungliga Svenska Vetenskapsakademiens. Handlingar. 1898;**24**:1-39. Available from: https://www.sid.ir/en/journal/ViewPaper.aspx?ID=263167 [Accessed: August 16, 2020]

[64] Wu F-C, Tseng R-L, Juang R-S. Characteristics of Elovich equation used for the analysis of adsorption kinetics in dye-chitosan systems. Chemical Engineering Journal. 2009;**150**(2):366-373. DOI: 10.1016/j.cej.2009.01.014

[65] Tseng R-L, Tran HN, Juang R-S. Revisiting temperature effect on the kinetics of liquid–phase adsorption by the Elovich equation: A simple tool for checking data reliability. Journal of the Taiwan Institute of Chemical Engineers. 2022;**136**:104403. DOI: 10.1016/j.jtice.2022.104403

[66] Wu F-C, Tseng R-L, Huang S-C, Juang R-S. Characteristics of pseudo-second-order kinetic model for liquid-phase adsorption: A mini-review. Chemical Engineering Journal. 2009;**151**(1):1-9. DOI: 10.1016/j.cej.2009.02.024

[67] Manzo-Valencia S, Alfaro-Cuevas-Villanueva R, Zárate-Medina J, Manzo-Valencia MK, Huirache-Acuña R, Cortés-Martínez R. Arsenate adsorption from aqueous solutions using chitosan/Fe-nanoparticles composite microspheres. ChemistrySelect. 2024;**9**(14):e202304662. DOI: 10.1002/slct.202304662

[68] Nieto-Delgado C, Gutiérrez-Martínez J, Rangel-Méndez JR. Modified activated carbon with interconnected fibrils of iron-oxyhydroxides using Mn2+ as morphology regulator, for a superior arsenic removal from water. Journal of Environmental Sciences. 2019;**76**:403-414. DOI: 10.1016/j.jes.2018.06.002

[69] Mansour R, Simeda MG, Zaatout A. Adsorption studies on brilliant green dye in aqueous solutions using activated carbon derived from guava seeds by chemical activation with phosphoric acid. Desalination and Water Treatment. 2020;**202**:396-409. DOI: 10.5004/dwt.2020.26147

[70] Debamita C, Rampal N, Gautham JP, Vairavel P. Process optimization, isotherm, kinetics, and thermodynamic studies for removal of remazol brilliant blue-R dye from contaminated water using adsorption on guava leaf powder. Desalination and Water Treatment. 2020;**185**:318-343. DOI: 10.5004/dwt.2020.25395

[71] Giles CH, Smith D, Huitson A. A general treatment and classification of the solute adsorption isotherm. I. Theoretical. Journal of Colloid and Interface Science. 1974;**47**(3):755-765. DOI: 10.1016/0021-9797(74)90252-5

[72] Chen C L, Park S W, Su JF, Yu Y-H, Heo J, Kim K, et al. The adsorption characteristics of fluoride on commercial activated carbon treated with quaternary ammonium salts (Quats). Science of The Total Environment. 2019;**693**:133605. DOI: 10.1016/j.scitotenv.2019.133605

[73] Pillai P, Lakhtaria Y, Dharaskar S, Khalid M. Synthesis, characterization, and application of iron oxyhydroxide coated with rice husk for fluoride removal from aqueous media. Environmental Science and Pollution Research. 2020;**27**(17):20606-20620. DOI: 10.1007/s11356-019-05948-8

[74] Pang T, Aye Chan TS, Jande YAC, Shen J. Removal of fluoride from water using activated carbon fibres modified with zirconium by a drop-coating method. Chemosphere. 2020;**255**:126950. DOI: 10.1016/j.chemosphere.2020.126950

[75] Shimelis B, Zewge F, Chandravanshi BS. Removal of excess fluoride from water by aluminum hydroxide. Bulletin of the Chemical Society of Ethiopia. 2006;**20**(1):17-34. DOI: 10.4314/bcse.v20i1.21140

[76] Yang W, Shi F, Jiang W, Chen Y, Zhang K, Jian S, et al. Outstanding fluoride removal from aqueous solution by a La-based adsorbent. RSC Advances. 2022;**12**(47):30522-30528. DOI: 10.1039/D2RA06284D

[77] Taibi Z, Bentaleb K, Bouberka Z, Pierlot C, Vandewalle M, Volkringer C, et al. Adsorption of Orange G dye on hydrophobic activated bentonite from aqueous solution. Crystals. 2023;**13**(2):211. DOI: 10.3390/cryst13020211

[78] Sánchez-Rodríguez S, Trujillo-Reyes J, Gutiérrez-Segura E, Solache-Ríos M, Colín-Cruz A. Removal of indigo carmine by a Ni nanoscale oxides/schoenoplectus acutus composite in batch and fixed bed column systems. Separation Science and Technology. 2015;**50**(11):1602-1610. DOI: 10.1080/01496395.2014.986579

[79] Alfaro-Cuevas-Villanueva R, Hidalgo-Vázquez AR, C De J CP, Cortés-Martínez R. Thermodynamic, Kinetic, and equilibrium parameters for the removal of lead and cadmium from aqueous solutions with calcium alginate beads. The Scientific World Journal. 2014;**2014**(1):647512. DOI: 10.1155/2014/647512

[80] Argun ME, Dursun S, Ozdemir C, Karatas M. Heavy metal adsorption by modified oak sawdust: Thermodynamics and kinetics. Journal of Hazardous Materials. 2007;**141**(1):77-85. DOI: 10.1016/j.jhazmat.2006.06.095

Chapter 4

Ion Adsorption and Hydration Superlubricity

Tianyi Han, Chenhui Zhang and Jianbin Luo

Abstract

The surface-water interactions are pivotal in interfacial functions and properties relevant in many natural processes, particularly the lubrication systems of living organisms, like the articular cartilage, oral mucosa, cardiovascular system, muscle, and tendon tissue. The ion adsorption often determines interfacial lubrication performance, especially hydration superlubricity, which depends on the surface charge characteristics and ion types. The hydration superlubricity mechanism has been well understood over the past 30 years, which further promotes the exploration and development of artificial hydration lubrication materials inspired by nature. This review first introduces the basic properties of the solid–liquid interface and the superlubricity system related to hydrated ions. Then, the influencing factors and mechanism of interfacial ion adsorption on hydration superlubricity are presented. Subsequently, negatively charged materials based on overall modification are summarized, including modified PEEK and modified UHMWPE with ultra-low friction coefficient and ultra-low wear rate. Finally, the potential application of such materials in artificial joints and underwater equipment systems are put forward. It is projected that interest in ion adsorption and hydration superlubricity from the basic science and engineering communities will continue to accelerate in the near future, accompanied by the concept of superlubricitive engineering.

Keywords: hydration lubrication, superlubricity, ion distribution, interfacial adsorption, surface force, surface charge density, zeta potential

1. Introduction

The ubiquity of aqueous electrolytes in contact with charged surfaces is relevant to a plenty of biological, environmental, physical and chemical processes, as well as potential applications like bioimplants, lubricants, batteries, and catalysis. A more common case is that charged surfaces interact with ions and water molecules in electrolyte solutions [1]. In animal fluids (like tears and joint fluids), ions (mainly Na^+, K^+, Mg^{2+}, Ca^{2+}, and Cl^-) exist at a concentration of ~0.2 M. This is roughly the same as the relative proportion of these ions in the ocean, but the total concentration (~0.6 M) is higher. Even pure water (pH = 7) can be considered as an electrolyte solution containing 10^{-7} M H^3O^+ and OH^- ions, which is sometimes very important. For example, hydrophobic polymers are negatively charged in pure water due to the orientation of adsorbed H_2O and OH^- [2, 3]. Until now, although we have gained a clear understanding of the electric double-layer structure [1], many open questions remain regarding the distributions and structures of water molecules and ions, the interfacial organization and competitive adsorption of water molecules and different ions, and preferential alignment of water molecules at different charged interfaces [4–11].

Besides, the adsorption of molecules to surfaces in liquids plays a significant role in lubrication, providing effective methods to solve friction and wear problems, which are accompanied by many aspects of production and life. Hardy et al. proposed boundary lubrication 100 years ago, where lubricants are adsorbed on friction surfaces through a physical or chemical process to form an adsorption film [12]. Luo et al. proposed thin film lubrication (TFL) 30 years ago, where the lubricant film substantially involves three layers, that is, the adsorbed layer, the ordered layer, and the fluid layer (**Figure 1a**), which fills the gap between boundary lubrication and elastohydrodynamic lubrication regimes [13, 14]. In electrolyte solutions, the adsorbed layer consists of hydrated counterions, which is named as the hydration layer and is significant for lubrication between charged surfaces [15, 16]. We can consider this hydration lubrication as boundary lubrication at high load regime or thin film regime (where the shear-promoted thermally activated sliding dominates the friction energy dissipation) and as TFL at low load regime or thick film regime (where the pathway of energy dissipation mainly results from the viscous shear) [6, 17, 18].

Given the important role of interfacial adsorption in lubrication, the main objective of this chapter is to discuss the influence mechanism of ion adsorption on hydration superlubricity and the idea for designing negatively charged modified materials based on this mechanism. Here, the basic properties of solid-liquid interface are first introduced, especially the electric double-layer model, which is the key to understand hydration lubrication. Then the phenomena and mechanisms of liquid superlubricity are summarized, especially the hydration superlubricity at nanoscale and macroscale conditions. In particular, we focus on how ion adsorption determines hydration superlubricity, especially the type, structure, and density of adsorbed ions. The influence of surface charge characteristics, surface charge density, ion valence, and ionic type on the ion adsorption is discussed. According to the mechanism of hydration superlubricity, we have proposed the idea for designing new polymer materials with higher surface charge density and better lubrication performance, that is, negatively

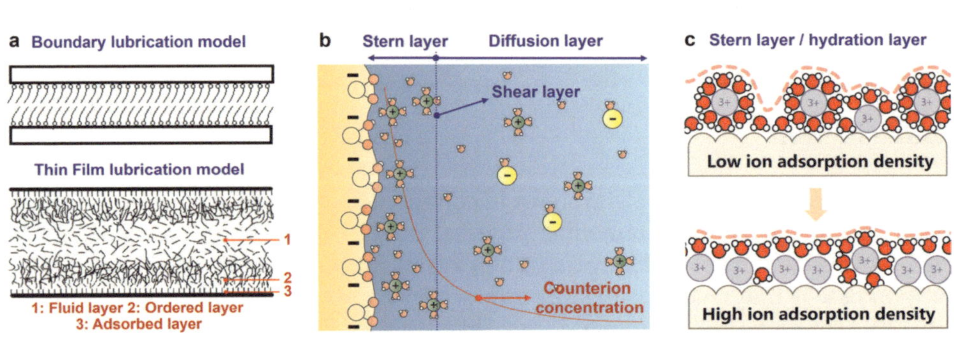

Figure 1.
Lubrication model and electrical double-layer (EDL) model at the solid-liquid interface. (a) Boundary lubrication model and thin film lubrication model. Reproduced with permission from Ref [13], Elsevier 1996. (b) EDL model, including the Stern layer and the diffusion layer. (c) Evolution of the Stern layer (also called hydration layer under nanoconfinement) structure with the La^{3+} *ion concentration. Reproduced with permission from Ref [6], Authors 2023.*

charged modified materials, which exhibit lower friction coefficient (<0.01) and lower wear rate ($<10^{-7}$ mm^3 (N·m)$^{-1}$). The understanding and study of hydration lubrication and hydrated lubricating materials would promote the research in applications of artificial joints (working in body fluids) and underwater equipment (working in oceans). We could expect a 'near-zero-friction' world based on hydration lubrication.

2. The solid-liquid interface

When a solid surface is in contact with an aqueous solution, it acquires a surface charge due to three major mechanisms, including the reaction of surface functional groups (acidic groups: such as-OH, -COOH, -SO_3H; basic groups: such as -NH_2), the adsorption of water ions (such as OH^- and H_3O^+), and the dissolution of ions (such as K^+) into the bulk. The charge formation of surfaces like Si_3N_4, Al_2O_3, SiO_2, etc., is dominated by the first process, while the second process dominates the surface charge of hydrophobic materials [2] such as polytetrafluoroethylene (PTFE), ultra-high molecular weight polyethylene (UHMWPE), single crystalline diamond (SCD) [19], etc. Surfaces like mica [20] and clinochlore [8] are charged due to the third mechanism. For the first and second mechanisms, the sign and density of surface charges depend on the pH of the aqueous solution, while the influence of pH is weakened for the surfaces dominated by the third mechanism [20]. No matter what charge formation is, the charged surface affects the ion distribution in the aqueous solution due to the electrostatic interaction. Counterions are attracted closer to the surface, while co-ions are repelled away from the surface, and thus, the electrical double layer (EDL) is generated, which is composed of the Stern layer (a film of the counterions closely attached to the charged surface by the electrostatic force) and the diffusion layer (a film of the dispersion solvent adjacent to the surface, containing free ions with a higher concentration of the counterions), schematically shown in **Figure 1b**. The thickness of the stationary Stern layer is usually below 1 nm, and the thickness of the diffusion layer is known to be related to the Debye length. When the liquid flows across the solid surface, the shear occurs between the stationary layer and the diffusion layer, which is known as the shear plane/layer. The zeta potential describes the charging behavior at the solid-liquid interface, which is usually considered as the potential at the shear plane and can be analyzed by electrokinetic methods [20–22], normal interaction measurements [6, 23], and molecular dynamics simulations [24, 25]. It should be pointed out that the EDL model assumes that (1) ions are point charges and only interact through Coulombic interactions; (2) there are no ion-ion correlations; (3) water is a homogeneous dielectric continuum [4]. Therefore, the EDL model is only suitable for aqueous solutions at relatively low concentrations, while it does not match numerous experiments at relatively high concentrations [26–29].

Compared with the diffusion layer, in which the potential and ionic density can be accurately described by the Poisson-Boltzmann (PB) equation, little is known for the distribution and structures of counterions in the Stern layer [1]. The interfacial potential in absolute value decays linearly within the Stern layer from the surface to the head end of the diffusion layer and subsequently decays further toward the bulk of solution [30]. The Stern layer thickness and capacitance and the amount of ordered/oriented/bound water in the Stern layer have been studied [31–33], but these parameters are currently only employed as rough estimates. The counterion adsorption behavior in the Stern layer as well as its influence factors are still not well understood, which need further research based on advances in experiment, computation, and theory. Furthermore, the Stern layer is also known as the hydration layer under nanoconfinement in the solid-liquid-solid system, especially in the field of lubrication [16, 34]. Water molecules form a tenacious hydration shell around charges/ions due to the considerable reduction in free energy arising from the reduction in the Born energy of the charge by having the water dipoles arranging around it [35, 36]. Therefore, the hydration layer is strongly attached and resists being removed or squeezed out by compression. For a long time, the structure of hydrated ion layers adsorbed on solid surfaces was thought to be invariable [16, 17, 37]. Recently, we prove that the ion surface coverage dictates the roughness of the hydration layer under subnanometer confinement, which depends on the bulk electrolyte concentration [6], as shown in **Figure 1c**. The subtle structural changes within the layers of adsorbed ions, both proximal (hemispherical) and distal (spherical), are influenced by the increasing concentration, resulting in redistribution of the hydration water. At low concentrations, the predominance of two layers of adsorbed ions creates a rough hydration layer (Rz ≈ 1.0 Å) on the surfaces, while at higher concentrations, the proximal hydrated ions increase and smoothen the hydration layer (Rz ≈ 0.3 Å). The change of the hydration layer structure plays a significant role in various natural and industrial processes including lubrication [6] (and hydration lubrication was selected as one of the 2024 top 10 emerging technologies in chemistry by International Union of Pure and Applied Chemistry (IUPAC) [38]), ion separation [39], and biological membranes [5], etc.

3. Hydration superlubricity

The characteristics of the solid-liquid interface determine the lubrication behavior. Liquid molecules with high load bearing capacity (strong intermolecular interactions) and low shear resistance (low intermolecular interactions) are considered to exhibit excellent lubrication performance between solid surfaces. The challenge is that the two are usually mutually exclusive. The state of ultralow friction (with friction coefficient at 10^{-3} level or less) under lubrication with liquids between two sliding surfaces is known as the liquid superlubricity [15, 40, 41], making near-zero friction and near-zero wear possible. In 1937, physicist P. Kapitza discovered the superfluidity of liquid helium (^{4}He), where the friction inside the fluid disappears at 2.17 K, namely, the viscosity of the fluid becomes zero, which is considered as the earliest phenomenon of liquid superlubricity. However, a liquid with zero viscosity cannot support applied loads based on the hydrodynamic lubrication theory, leading to lubrication failure. According to Reynolds equation, increasing the viscosity of the liquid lubricant can improve the load-carrying capacity, but will also result in higher shear resistance and friction, and vice versa. This inherent contradiction leads to a lower limit of the friction coefficient, which cannot be further reduced. How to break through the lower limit of friction coefficient and achieve superlubricity? From the sources of friction, the shear resistance (F_s) arises from two interactions: one is between the surface roughnesses ($F_{roughness}$), and the other is within the fluid film (F_{fluid}), where $F_s = F_{roughness} + F_{fluid}$. The shear force between surface roughnesses is usually much higher than that within the fluid film, $F_{roughness} \gg F_{fluid}$. Therefore, reducing $F_{roughness}$ is the key to break through the lower limit and further achieve superlubricity. However, which interactions can be used to separate the two surfaces (high load-carrying capacity) and obtain ultralow F_s (low shear resistance)? This is a significant challenge.

Over the past thirty years, great progress has been made in the research of liquid superlubricity [15, 40–45]. Luo and Zhang's group summarized five major types of liquid superlubricity (**Tables 1**, 15], identified three key mechanisms for achieving liquid superlubricity (hydrodynamic effect, electric double layer and hydration effect, and tribo-induced films, as shown in **Figure 2**), and first proposed the concept of 'superlubritive engineering' to turn superlubricity from laboratory research into engineering applications [40]. At present, the minimum friction coefficient and the maximum contact pressure for liquid superlubricity are in the order of 10^{-5}–10^{-4}, and over 1 GPa, respectively. As one of the frontier subjects in tribology, more superlubritive materials are being explored, the atomic-scale mechanisms of superlubricity are being revealed, and the applications of superlubricity are being promoted [15, 54–56].

Next, we mainly focus on the hydration superlubricity based on the adsorption of hydrated ions. 40 years ago, the short-range repulsive force was observed by Israelachvili and Pashley in electrolyte solutions between two atomically smooth mica surfaces through a surface force apparatus, which is known as the hydration repulsion [57–60]. About 30 years ago, Ducker et al. measured the normal forces between two silica surfaces in NaCl solutions using an atomic force microscope and also found the short-range hydration force [61, 62]. The hydration force can support applied load

Lubricant materials		COF	Pressure/MPa	Methods
Water, acid-based, alcohol-based solutions		0.002–0.009	$5 \sim 200$	Tribometer
Hydrated materials	Hydrated ions	$10^{-4} - 0.01$	10, $270 \sim 550$	SFA/SFB, Tribometer
	Phospholipids	$10^{-5} - 0.002$	40	SFA/SFB
	Polymer brushes	$10^{-4} - 0.01$	15	SFA/SFB, Tribometer
	Surfactants	$10^{-4} - 0.01$	10	SFA/SFB, AFM
	Hydrogels	0.001–0.01	1	Tribometer
	Negatively charged modified materials[b]	0.005–0.02	110	Tribometer
Ionic liquids		0.001–0.006	130	AFM, SFA/SFB, Tribometer
Solid-liquid combined superlubricants[c]		0.0006–0.006	1400	Tribometer
Oil-based lubricants		0.001–0.008	100	Tribometer

[a]This table is based on our previous review article [15]. SFA and SFB mean surface force apparatus and balance, respectively, and AFM means atomic force microscope.[b]Negatively charged modified materials is a new concept based on hydration contribution. Various new materials have been designed, such as carboxylate anions ($RCOO^-$)-grafted epoxy resins [46], sulfonated PEEK [47], and polyelectrolyte-grafted UHMWPE [48, 49]. More details will be introduced in Part 5.[c]It is also known as the 2D materials as lubricant additives. The solid-liquid combined superlubricity can also be considered as a separate classification independent of liquid superlubricity due to the unique role of solid materials. The contact pressure during superlubricity was above 1.4 GPa using molybdenum carbide (MXene) and ionic liquids or lithium citrate solution as the lubricant [50, 51].

Table 1.
Five different categories of liquid superlubricity systems[a].

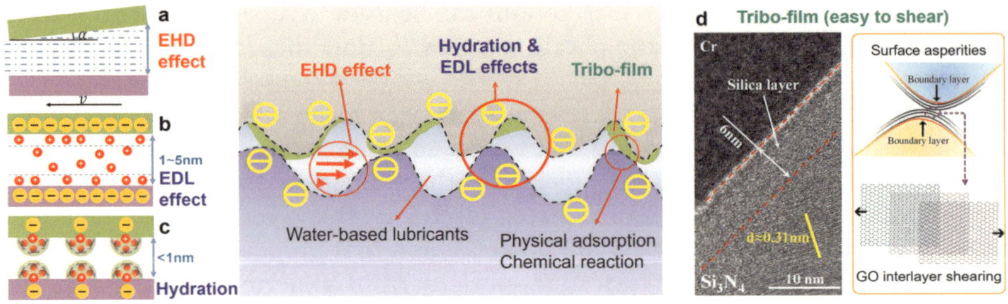

Figure 2.
Liquid superlubricity mechanisms. (a) Elastohydrodynamic (EHD) effect. (b) EDL effect. (c) Hydration effect. (d) Tribo-film which is easy to shear. Reproduced with permission from Ref [52], American Chemical Society 2019. Reproduced with permission from Ref [53], Elsevier 2019.

and prevent the direct contact of surfaces, which is expected to resolve the inherent contradiction to achieve superlubricity. Based on the hydration repulsion generated by the hydration layer (**Figures 3a** and **2c**), ultralow friction coefficient of 0.0002–0.002 was obtained between two mica surfaces in 0.1 M Na^+ solutions (**Figure 3b**) [17]. Klein's group attributed this to the ready exchange (as opposed to loss) of water molecules within the hydration layers as they rubbing past each other under strong compression [16, 65]. Meanwhile, Higashitani's group investigated the normal and friction forces between two silica surfaces in electrolyte solutions (**Figure 3c-e**) and attributed the lubrication property to the hydration capacity of cations, where more hydrated the cations are, the stronger the lubrication is [63]. Furthermore, different pathways of friction energy dissipation for hydration lubrication of monovalent and multivalent cations were proposed based on the velocity-dependent friction. At low pressure and large film thickness, the viscous shear of fluid film dominates, where a power law relationship describes the velocity dependence of the friction force ($F_s \sim A\eta(v/D)^n$ based on a Newtonian relation); while at high pressure and small film thickness, the shear-promoted thermally activated sliding plays a pivotal role, where a logarithmic relation exists between the friction force and velocity ($F_s \sim \ln(v)$ based on the Eyring theory) [6, 17, 18]. Besides, atomic-scale three-dimensional hydration structures were visualized on charged surfaces (like clinochlore, mica, and calcite) by using ultra-low noise frequency-modulation atomic force microscopy (FM-AFM) [7, 8, 66].

In the past decade, Luo and Zhang's group has constructed the macroscale hydration superlubricity system with high load-carrying capacity by introducing microscopic surface forces (including hydration force and electric double-layer force) into macroscale contact [15]. The Reynolds equation considering surface forces (P_{sf}) was established [67–69], as shown in Eq. (1–4). The bearing ratio of surface forces can reach over 80%, [15]. In other words, the surface force can bear most of the applied pressure (P) at the boundary and mixed lubrication regimes, where the hydrodynamic effect (P_{hd}) is weak.

$$\frac{\partial}{\partial x}\left(\frac{\rho h^3}{12\eta}\frac{\partial P_{hd}}{\partial x}\right)+\frac{\partial}{\partial x}\left(\frac{\rho h^3}{12\eta}\frac{\partial P_{hd}}{\partial x}\right)=U\frac{\partial(\rho h)}{\partial x} \tag{1}$$

$$h=h_0+s(x,y)+\frac{2}{\pi E'}\iint_\Omega \frac{P}{\sqrt{(x-\xi)^2+(y-\eta)^2}}d\xi d\eta \tag{2}$$

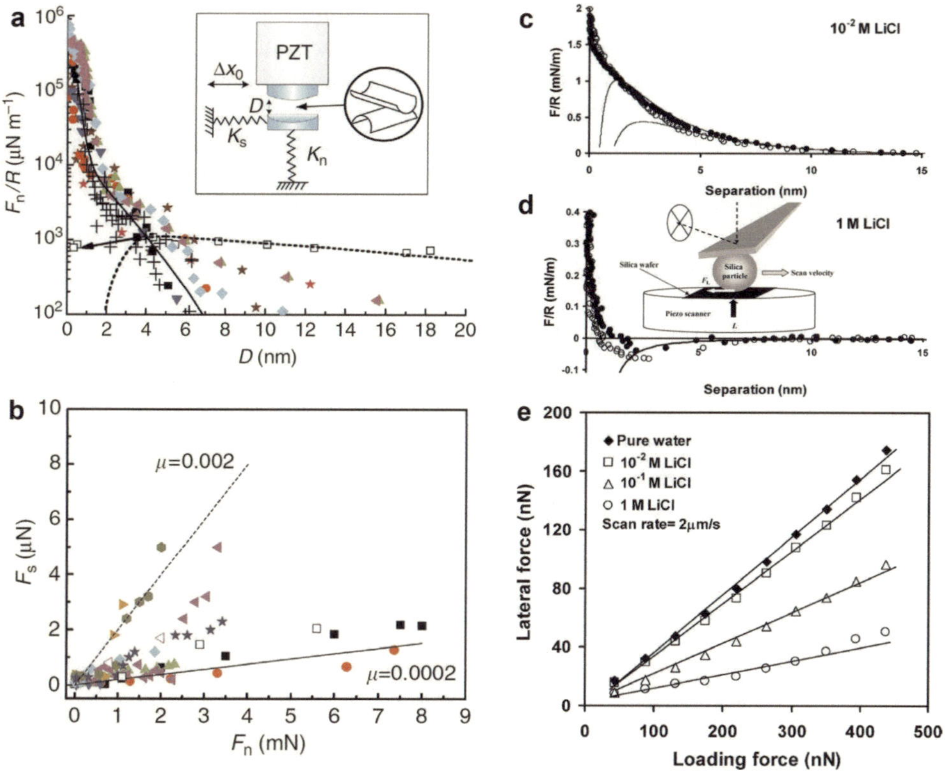

Figure 3.
Hydration lubrication. SFB measurements of (a) normal forces and (b) shear forces across 0.1 M NaCl and $NaNO_3$ between two mica surfaces. Reproduced with permission from Ref [17], Authors 2015. The inset in (a) shows the schematic SFB configuration. AFM measurements of normal forces at (c) 0.01 M and (d) 1 M LiCl, and (e) shear forces between two silica surfaces. Reproduced with permission from Ref [63], American Chemical Society 2005. The illustration in (d) shows the schematic AFM configuration. Reproduced with permission from [64], Elsevier 2006.

$$P = P_{sf} + P_{hd} + P_{sr} \tag{3}$$

$$P_{sf} = P_{hf} + P_{dl} + P_{vdW} \tag{4}$$

where h means the film thickness between two surfaces, including three contributions of rigid body displacement, surface morphology, and elastic deformation; P_{sf} means the pressure supported by surface forces, P_{hd} means the hydrodynamic contribution, P_{sr} means the pressure supported by surface roughness, and P_{hf}, P_{dl}, and P_{vdW} represent the hydration force, electric double-layer force, and van der Waals attraction force, respectively.

In part 2, we learned that most surfaces bear an electric charge in water (**Figure 1b**). If two charged surfaces approach each other and the EDLs overlap (**Figure 2b**), the electric double-layer force arises due to the entropic (osmotic), not electrostatic contribution. The repulsive osmotic pressure between the counterions maintains the EDL, which forces them away from the surface and from each other so as to increase their configurational entropy [70]. When two charged surfaces are brought close to each other, the counterions are compelled to return to the surfaces, against their preferred equilibrium state, which is resistant to the osmotic repulsion but beneficial for the electrostatic attraction. The osmotic repulsion dominates, and thus, the net force is repulsive. Formulation of electric double-layer force is shown in

Eq. (5) [67]. Regarding the origin of hydration repulsion [1, 35], it is a type of solvation force, which is essentially an ion–dipole interaction. One ion can form multiple ion-dipole bonds, resulting in the ion being solvated and orientationally bound by solvent molecules to form solvated ions (also known as hydrated ions in water). The solvated ions are bound strongly and are restructured by the polar surfaces. An ordered solvent layer was formed at the surface-solution interface, which quasi-exponentially decays away from the surface. An overlap of the ordered-solvent layers near the two mutually approaching surfaces creates a repulsive force, that is, hydration force (**Figures 2c, 3a, c, d**). Its formulation is expressed as Eq. (6) [67].

$$P_{\mathrm{dl}} = 64ck_{\mathrm{B}}T\tanh(e\psi_0/4k_{\mathrm{B}}T)e^{-\kappa h} \tag{5}$$

$$P_{\mathrm{hf}} = P_0 e^{-h/\lambda} \tag{6}$$

where c is the ionic concentration, k_{B} is Boltzmann's constant, T is the absolute temperature, ψ_0 is the surface potential, κ is the Debye length, P_0 is the hydration force amplitude, and λ is the decay length of hydration repulsion.

Guided by this theory Eq. (1–6), we designed an experiment by using a tribometer as follows. First, an acid solution with pH = 1.5 was used to lubricate the Si_3N_4/sapphire friction pairs for 5 minutes, in order to increase the charge density of friction surfaces; afterwards, the acid solution was removed and replaced with the electrolyte (LiCl, NaCl, and KCl) solution, and then, the friction coefficient immediately decreased to 0.005 (**Figure 4a**), achieving macroscale superlubricity under high contact pressures above

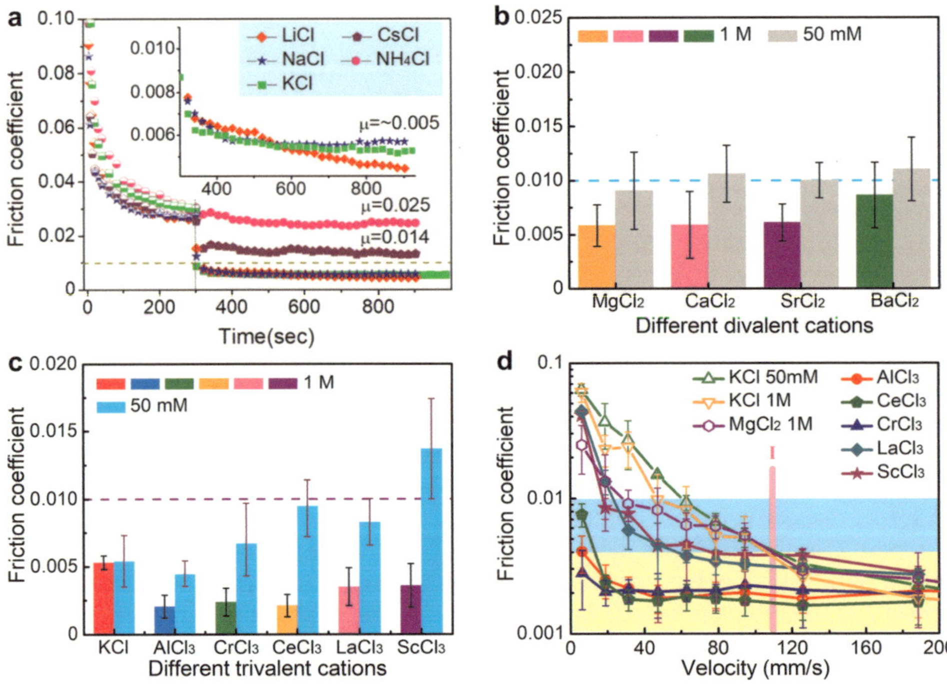

Figure 4.
Macroscale hydration superlubricity. The friction pairs are a Si_3N_4 ball and a sapphire disk. (a) Friction coefficients of five monovalent chloride solutions (50 mM) after running-in with acid solutions. Reproduced with permission from [71], American Chemical Society 2018. Friction coefficients of (b) four divalent and (c) five trivalent chloride solutions. (d) Velocity-dependent friction coefficients. The concentration of five trivalent solutions was 1 M. Reproduced with permission from [72], American Chemical Society 2020.

250 MPa [52, 71]. The superlubricity mechanism was mainly attributed to the hydration effect. First, both surfaces are negatively charged in neutral electrolyte solutions, which can adsorb counterions to form the hydration layer [71]. Second, a silica layer was generated on Si_3N_4 surface during the acid running-in period, which increased the surface charge density by more than 50% and thus adsorbed more hydrated cations [52]. Furthermore, we evaluated the hydration lubrication performance of multivalent cations between Si_3N_4/sapphire surfaces [72]. Before that, the normal interactions between two mica surfaces in divalent (Mg^{2+}, Ca^{2+}, Sr^{2+}, Ba^{2+}, and Ni^{2+}), trivalent (La^{3+}, Cr^{3+}), and tetravalent (Th^{4+}) electrolyte solutions were studied based on surface force measurements [57, 73–80]. At higher concentrations ($\geq$ 0.1–1 M), both divalent and trivalent cations gave rise to strong, short-range, repulsive force due to the hydration shells of the bound cations [73, 74, 79]. Therefore, it is reasonable to speculate that multivalent cations can also exhibit hydration lubrication. Based on our experiments using a tribometer, both divalent and trivalent cations studied achieved superlubricity with the lowest friction coefficient of 0.002–0.005 at 1 M under a high contact pressure above 270 MPa, but the friction coefficient all increased to a higher value at 50 mM (**Figure 4b-c**). For monovalent KCl solution, almost the same friction coefficient of 0.005 was obtained at both high and low concentrations. According to **Figure 1b**, the counterion adsorption density and the hydration layer structure decide the hydration lubrication performance, which may result in the difference of lubrication performance between monovalent and multivalent cations. More details will be introduced in Part 4.

Furthermore, the superlubricity capacity of different cations wase compared based on the velocity-dependent friction coefficient (**Figure 4d**). At higher velocities ($v > v_I \approx 110$ mm/s), all these cations and pure water achieved superlubricity with friction coefficients below 0.004 under contact pressures about 250 MPa, indicating that excellent superlubricity performance can be achieved solely through hydrodynamic effect, and the pressures were equivalent to the theoretical limit of the bearing capacity of hydrodynamic effect [81, 82]. However, at lower velocities ($v < v_I$), where the hydrodynamic contribution became weaker while hydration contribution became stronger based on Eq. (3), it is evident that trivalent cations exhibited the best superlubricity performance compared with divalent and monovalent cations. Three trivalent cations (Al^{3+}, Ce^{3+}, and Cr^{3+}) can even obtain superlubricity at very low velocity of 1–3 mm/s, showing that hydration superlubricity can be obtained at the boundary lubrication regime under high contact pressures above 250 MPa. Meanwhile, by introducing polyethylene glycol (PEG) or poly vinyl alcohol (PVA) into electrolyte solutions to enhance the hydrodynamic effect, the liquid superlubricity with ultralow friction coefficient about 0.004 and ultralow wear rate below 1.9×10^{-9} $mm^3/(N{\cdot}m)$ was achieved under extremely high contact pressures above 550 MPa [83, 84]. Therefore, macroscale superlubricity can be achieved under high pressures above 250 MPa only based on hydration contribution, while superlubricity under higher pressures above 550 MPa can be achieved based on the synergistic contribution of hydration and hydrodynamic effect, which extended the hydration lubrication concept from nanoscale contact at physiological pressures to more engineering-like pressures at macroscale contact [40, 42, 56, 65, 85].

4. Ion adsorption determines hydration lubrication

It is well understood that hydrated ions can achieve superlubricity between mica/mica [6, 16, 17, 37], ceramic/ceramic [52, 71, 72], mica/hydrophobic fluoropolymer [86], silica/calcite [18], mica/silicon [87], ceramic/ice [88, 89], and GCr15/ice [89] surfaces. We will further discuss a more fundamental question: how does the type of adsorbed ions affect the hydration lubrication? First, the influence of ion valence on hydration lubrication has been studied, showing that the larger the ion valence, the stronger the hydration strength of hydrated ions, the better the hydration lubrication performance. Besides, the superlubricity of monovalent cations can be obtained at both low and high concentrations; while for multivalent cations, higher concentration results in better lubrication performance. Second, regarding the influence of ionic charge, it was initially believed that only cations could achieve hydration superlubricity due to the weaker hydration strength of anions. For example, the hydration superlubricity between ceramic surfaces failed at pH below the isoelectric points (IEP) because both the surfaces were positively charged, and thus, mainly anions rather than cations were adsorbed on surfaces in this instance [52]. Recently, macroscale superlubricity of anions was also obtained between polymer/sapphire surfaces at acidic solutions with pH < IEP [3]. Strongly hydrated anions (like SO_4^{2-}) exhibited stable superlubricity with friction coefficient of 0.003–0.007 comparable to cations. Meanwhile, nanoscale superlubricity was also obtained in neutral electrolyte solutions on positively charged $Mg(OH)_2$ surface due to the role of hydrated anions (Cr^-, Br^-, I^-) [90]. The advantage of cations over anions is that cations have a larger load-carrying capacity, because anions are usually unable to achieve superlubricity at applied pressures above 50 MPa [3].

Moreover, the influence of adsorbed ions on hydration lubrication has also been confirmed in previous studies about electric potential controlled lubrication [91–96]. The adsorption/desorption of sodium dodecyl sulfate (SDS) surfactant affecting the friction of stainless steel, metals, and ceramics was studied under different potentials, showing lower friction at positive surface potentials due to DS^- anions adsorbed on the friction surface [97–100]. Potential controlled friction behavior in the finger/titanium alloy interface was found in aqueous electrolyte solutions, which was attributed to the adsorption of OH^-, bubbles effect, and formation of passive film [101]. The hydration lubrication was also 'switched' on or off in situ when an electrolyte solution was used to lubricate mica-gold interface [102]. Furthermore, the hydration superlubricity was modulated by electric potential in aqueous ionic solutions between silica/gold [103], silica/graphite [104], silicon/steel, and silicon/gold [105] surfaces with friction coefficients transforming between ~0.005 and ~ 0.035, between ~0.003 and 0.004–0.006, and between 0.001–0.009 and 0.312–3.400, respectively, which were all attributed to the adsorption of different hydrated ions. However, using electrochemical atomic force microscope (EC-AFM), an opposite trend of normal and shear interactions with applied electric potential was found in monovalent and trivalent electrolyte solutions between silica/graphite surfaces [23], as shown in **Figure 5**. In 1 mM NaCl solution, the negative applied potential reduced the friction coefficient (**Figure 5c**) by increasing the cation adsorption (**Figure 5a**); while in 1 mM $LaCl_3$ solution, the positive applied potential reduced the friction coefficient (**Figure 5d**) due to the adsorbed anions on graphite weakening the ion-ion correlation between trivalent ions (**Figure 5b**). Therefore, the spatial correlations between multivalent ions [79, 106] should be considered in hydration lubrication.

As shown in **Figure 1b**, the ion surface coverage dictates the 'roughness' of the hydration layer and its lubricating properties under subnanometer confinement, and a higher ion surface coverage is more conducive to superlubricity [6]. The mica with high charge density (2 e/nm^2) is usually chosen as one of the typical surfaces for studying hydration lubrication [16, 17, 37]. Besides, the acid running-in period is necessary during the macroscale hydration superlubricity between ceramic surfaces

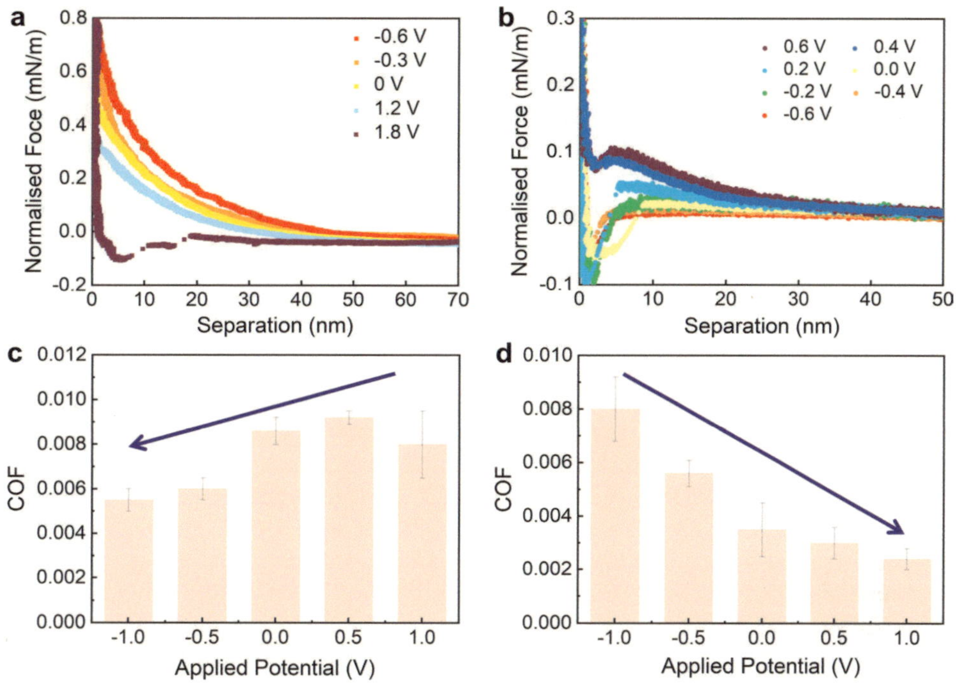

Figure 5.
Potential controlled hydration lubrication between SiO_2 and charged highly oriented pyrolytic graphite (HOPG) surfaces. (a) Normal forces and (c) friction coefficients in 1 mM NaCl solution. (b) Normal forces and (d) friction coefficients in 1 mM $LaCl_3$ solution. Reproduced with permission from [23], American Chemical Society 2024.

because the tribochemical process leads to a larger negative potential and a higher ion surface coverage on the friction surface [52, 71]. The ion adsorption density or surface coverage has been invoked to account for the hydration lubrication performance, but not well understood. A method to evaluate the adsorption density or surface coverage of counterions on a charged surface was proposed based on the streaming current measurements (**Figure 6a**) and expanded Grahame equation [20], as shown in **Figure 6**. It is surprising that more than 90% of charge sites on the negatively charged surfaces were neutralized by the counterions in electrolyte solutions with low concentration of 1 mM, but the coverage of trivalent cations (~1/3) or divalent cations (~1/2) was obviously lower than that of monovalent cations (>90%) (**Figure 6b**). Therefore, trivalent and divalent cations at lower concentrations of 50 mM exhibited a larger friction coefficient than monovalent cations (**Figure 4b, c**). Besides, the charge inversion existed in trivalent electrolytes (**Figure 6c, d**), which has been found in previous experiments [79, 107, 108], but the inversion concentration was predominantly affected by the ion type rather than the charged surface (**Figure 6c**). It is precisely because of the excessive adsorption that enough trivalent cations can be adsorbed on friction surfaces, thereby achieving superlubricity [6, 72]. As shown in **Figure 6c**, the charge inversion occurred at concentrations between 10^{-3} and 10^{-2} mM and between 0.1 and 1 mM in Al^{3+} and La^{3+} solution, respectively, for all three surfaces (mica, sapphire, and Si_3N_4), indicating that the counterion adsorption on charged surfaces was mainly attributed to the ionic competitive adsorption and the ion-ion correlation [20, 106]. However, based on the theoretical calculation modified by spatial correlations between multivalent ions at the surface, the charge inversion concentration was reduced with the dielectric constant decreasing and the bare surface charge density increasing [109]. Further studies are necessary to determine the importance of ion adsorption on charge inversion. Molecular dynamics simulations of solid–liquid interfaces provide molecular insights into local electric fields and ion adsorption [24, 33, 110–112], which may help us to understand the decisive role of ion adsorption to hydration lubrication.

According to the surface coverage (**Figure 6b**) and the lubrication performance (**Figure 4b** and **c**) of cations, insufficient adsorption of hydrated ions leads to superlubricity failure or increased friction coefficient, but the mechanism is still unclear. In order to clarify the influence of ion adsorption, we further measured the normal interactions in La $(NO_3)_3$ solutions between two mica surfaces using SFA (**Figure 7a, b**). At 25 mM, the sudden jump-in of the surfaces occurred at separation distance about 2 nm due to the weak adsorption of La^{3+} ions on mica surfaces and the squeezing out of these weakly adsorbed ions; while at higher concentrations of 100 mM to 2 M, strong hydration forces existed due to the strongly bound hydration layer. As expected, the weak ion adsorption at 25 mM resulted in surface damage during friction at very low load below 1 mN, while excellent hydration lubrication was obtained at higher concentrations without any mica damage (**Figure 7c**). Due to the inaccurate measurement of zeta potential at high concentrations above 100 mM based on the streaming current method (**Figure 6**), high-resolution atomic force microscopy (HR-AFM) was used to

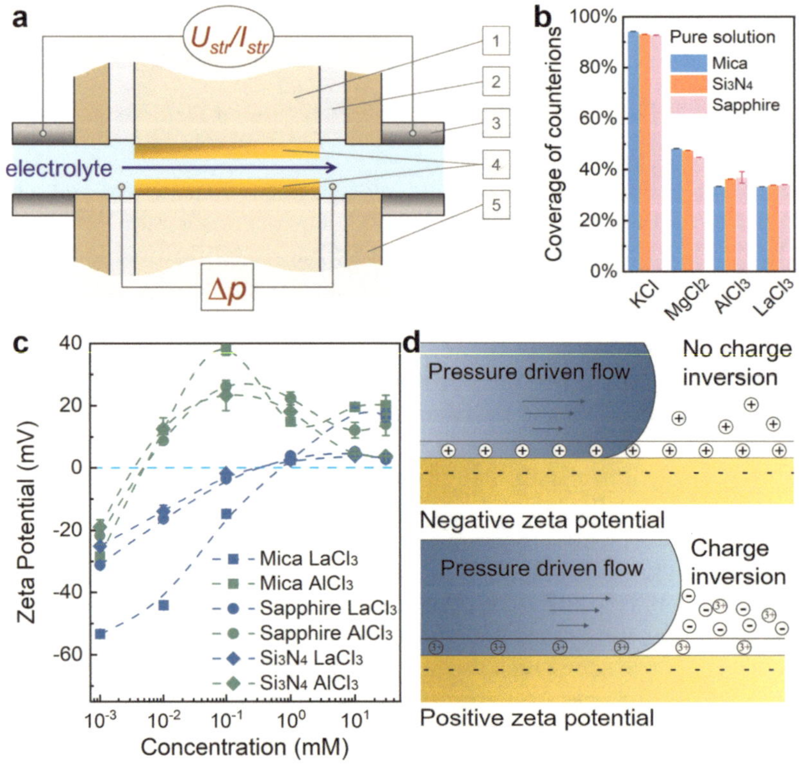

Figure 6.
Surface coverage of adsorbed ions and charge inversion. (a) Zeta potential measurements based on streaming potential/current measurements. (b) Surface coverage of cations in 1 mM electrolyte solution. (c) Concentration dependent zeta potential in trivalent electrolyte solutions. (d) Schematic illustration with or without charge inversion. Reproduced with permission from [20], American Chemical Society 2024.

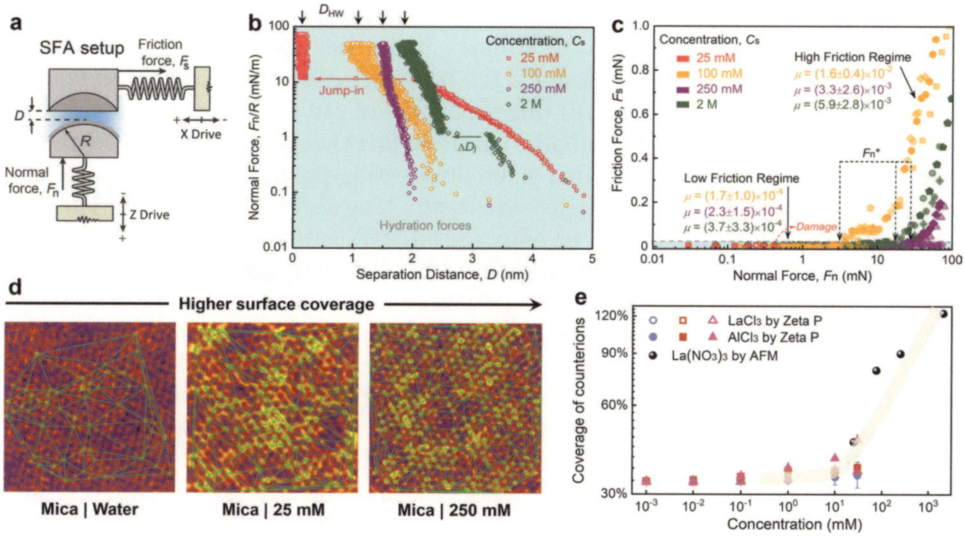

Figure 7.
Ion adsorption-dependent hydration layer structure modulates hydration superlubricity. (a) SFA setup. (b) Normal force and (c) friction force measurements between two curved mica surfaces across $La(NO_3)_3$ solutions. (d) HR-AFM topography of cleaved muscovite mica surfaces in water and $La(NO_3)_3$ solutions. (e) Surface coverage of trivalent cations on mica (blue and black symbols) and ceramic (Si_3N_4, red symbols; sapphire, magenta symbols) surfaces in electrolyte solutions ranging from 10^{-3} to 10^3 mM. 100% means that each surface charge site adsorbs a counterion. Reproduced with permission from [6], Authors 2023.

characterize the interfacial adsorption of La^{3+} ions on mica surface (**Figure 7d**), although an atomically smooth and contamination free surface is necessary for HR-AFM, which brings a big challenge to sample preparation. The ion surface coverage increased from ~0.9 ion/nm^2 (corresponding to a coverage of ~45%) to ~1.8 ions/nm^2 (corresponding to a coverage of ~90%) with the concentration increasing from 25 mM to 250 mM and started to form the second layer of ions as the concentration increased even further [6]. For mica in pure water, some hydrated K^+ ions adsorbed on the surface were identified due to the hydration and dissolution of potassium ions at the freshly cleaved interface into the aqueous media. Combined the AFM images and zeta potential measurements, the surface coverage of trivalent cations on mica and ceramic surfaces at a wide concentration range from 10^{-3} to 10^3 mM was obtained, as shown in **Figure 7e**. The ion coverage remained low (about 30–40%) at concentrations from 10^{-3} to 10 mM, while started to increase obviously at concentrations larger than 10 mM. It should be noticed that similar coverage for mica and ceramic surfaces does not mean similar amount of adsorbed ions due to a larger charge density for mica. More cations are adsorbed on the mica surface compared with ceramic surfaces at same concentration, but both surfaces show similar hydration superlubricity performance. For example, at high concentrations (~1 M), COF = 0.003–0.006 between mica surfaces by nanoscale SFA and COF = 0.003–0.004 between ceramic surfaces by macroscale UMT [72].

5. Negatively charged modified materials

Based on the understanding of hydration lubrication and interfacial ion adsorption, especially the realization of macroscopic hydration superlubricity, we have proposed an overall modification theory for designing superlubricating materials by increasing the surface negative charge density to enhance hydration lubrication [15, 46]. For surface modification, the lubrication will fail if modified layer is worn away; while overall modification perfectly solves this problem and can achieve continuous lubrication even if wear occurs [47, 49, 113]. Based on this theory, we have developed a series of negatively charged modified water-lubricating polymer materials with high load-bearing capacity, low friction, low wear, and long service life characteristics, including carboxylate anions-grafted epoxy resin [46], sulfonated modified polyether ether ketone (SPEEK) [47] and filler-reinforced PEEK composites [114, 115], polyelectrolyte-embedded modified UHMWPE [48, 49, 116, 117], etc. PEEK is a high-performance engineering plastic which possesses the advantages of high elastic modulus, anti-wear, hydrolysis resistance, dielectric property, thermal stability, and low water absorption. However, high friction coefficient (due to insufficient load-carrying capacity of water) causes heat and wear in PEEK product for water-lubricating applications. SPEEK with negatively charged sulfonic acid groups ($-SO_3^-$) has been developed [47], exhibiting lower surface potential and lower isoelectric point (compared with PEEK), an ultra-low friction coefficient lower than 0.02, and a super-low wear rate at a level of 10^{-8} mm^3 $(N{\cdot}m)^{-1}$, as shown in **Figure 8**. It should be noticed that PEEK powders were sulfonated using chlorosulfuric acid ($ClSO_3H$) and sintered to form bulk plastic, which ensures overall modification rather than just surface modification. SPEEK prepared by 11 hours of sulfonation reaction has a lower IEP of ~3.6, indicating that SPEEK possesses more negative charges in the solid/liquid interface, which can adsorb more hydrated cations, thus improving the hydration lubrication. Compared with the wear track on PEEK, SPEEK had extremely slight wear (**Figure 8b**). Furthermore, when SPEEK was compounded with 10 wt% carbon fiber, the friction coefficient decreased to 0.009 with the lubrication in 3 wt% NaCl solution, which was in the superlubricity regime, indicating the great potential to achieve superlubricity through the synergistic effect of hydration modification and fiber reinforcement [115].

UHMWPE is a polymer material with low friction, fine biocompatibility, and chemical stability. Surface grafting of polyethylene (PE) has been proven efficient in friction and wear reduction [118]. However, the surface-attached polymer brushes without a deep attachment are liable to be sheared off from the substrate with increasing sliding cycles, which results in lubrication failure and triggering the generation of wear particles. Embedding negatively charged polymer brushes uniformly inside the bulk UHMWPE suggests a more promising solution of improving lubrication longevity [48]. The hydrophilic 3-sulfopropyl methacrylate potassium salt (SPMK) monomer was chemically grafted onto UHMWPE powders through photoinduced radical graft polymerization (**Figure 9a**), followed by hot-press molding of the modified powders to afford homogeneous modified bulk material [48]. This polyelectrolyte-embedded UHMWPE exhibited an ultra-low friction coefficient below 0.01 and an ultra-low wear rate below 10^{-7} mm^3 $(N{\cdot}m)^{-1}$ under physiological conditions due to enhanced hydration contribution. To further enhance material strength and water resistance, SPMK functionalized carbon nanotubes (CNTs, **Figure 9b**) were used as the reinforcing agent to improve the mechanical and tribological properties of polyelectrolyte-embedded UHMWPE (**Figure 9c**). Modified UHMWPE composite had a lower zeta potential and a lower IEP (**Figure 9d**), thus adsorbing more hydrated cations on the surface. Modified UHMWPE composite reinforced by modified CNTs with a high aspect ratio of 1000 showed the greatest improvements in mechanical and water-resistant properties, which maintained

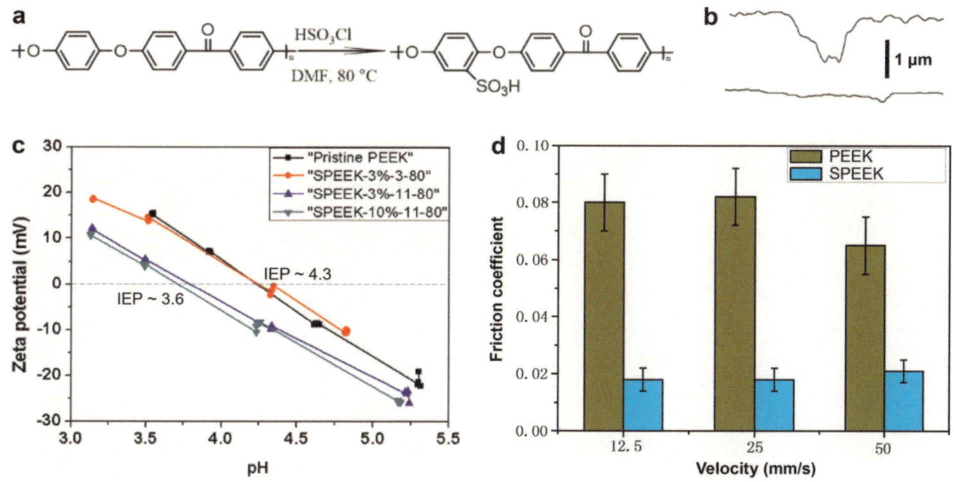

Figure 8.
Sulfonated PEEK (SPEEK) with excellent hydration lubrication performance. (a) Sulfonation procedure of the PEEK powders. (b) Cross-section of the wear tracks of PEEK (top) and SPEEK (down). (c) Zeta potential and (d) friction coefficient of PEEK and SPEEK. SPEEK-3%-3–80 in (c) represents the sample prepared according to (a), where 10% means the concentration of chlorosulfonic acid (ClSO$_3$H), 3 means the reaction time in h, and 80 means the reaction temperature in °C. Reproduced with permission from [47], Authors 2022.

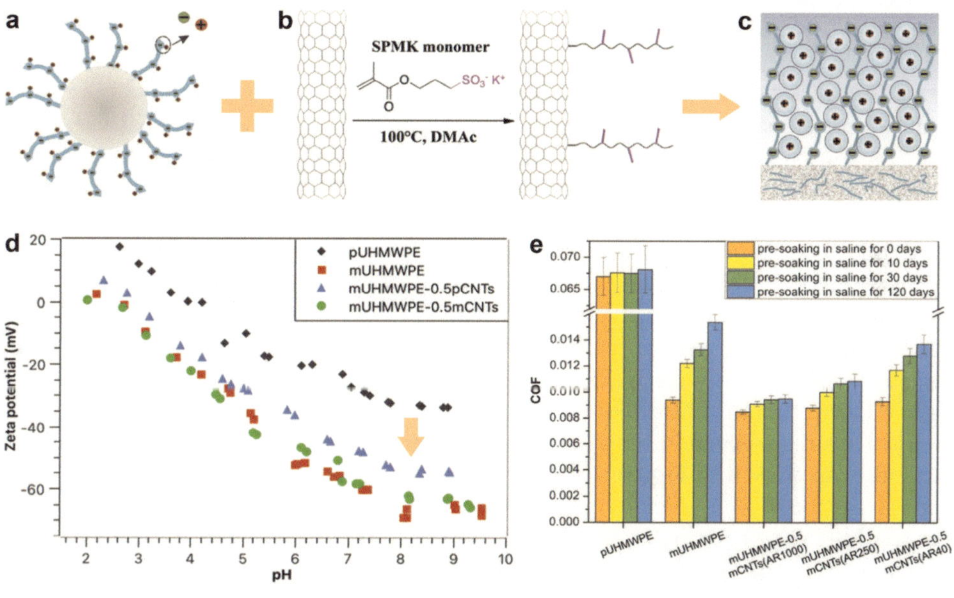

Figure 9.
Polyelectrolyte-embedded UHMWPE with functionalized carbon nanotubes reinforcement. (a) Modified UHMWPE powders. (b) Surface modification of CNTs through free-radical polymerization of SPMK monomer for the formation of surface-grafted polymer brushes. (c) Polyelectrolyte-grafted UHMWPE. Reproduced with permission from [48], American Chemical Society. (d) Surface zeta potential of UHMWPE with different compositions. (e) Friction coefficient of various UHMWPE composites after pre-soaking in saline. Reproduced with permission from [49], Elsevier 2024.

relatively stable superlubricity performance during 120-day saline pre-soaking, as shown in **Figure 9e**. This combination of strong hydration lubrication and superior material strength offers a promising design of superlubricating materials with low

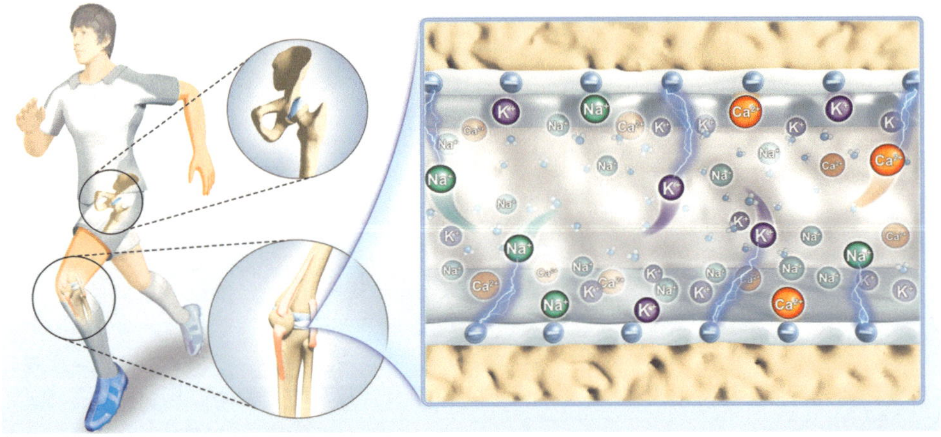

Figure 10.
Hydration superlubricity in biological systems such as synovial fluid lubricating joints (like hip joints and knee joints). The right illustration mainly shows the role of electrolytes (without proteins, polysaccharides, and hyaluronic acids) on hydration lubrication, indicating that significantly negatively charged surfaces are necessary.

friction, high wear resistance, and long lifetime. Besides, the carbon fiber and nano SiO_2 were filled into UHMWPE-grafted SPMK to further enhance its mechanical and tribological performance [117].

This idea of negatively charged modified materials depends on the hydration lubrication, which is a particular phenomenon observed in hydrated ions, polymer brushes, liposomes, surfactants, and other charged molecules, trapped between surfaces [15, 56, 65, 119]. For these overall modified materials with high surface charge density, hydration layers are adsorbed strongly on the surfaces due to electrostatic interactions, which generate hydration repulsion to sustain applied pressure, and meanwhile provide an impressive fluidity between hydration layers, leading to extremely efficient lubrication [16, 85, 119]. The exact mechanism is relatively clear, where the viscosity of hydration layers increases under nanoconfinement [6, 17, 18], hydrated ions seem to become 'solid-like' with the pressure further increasing, and thus, the friction energy dissipation depends on the 'roughness' of the hydration layer (**Figure 1b**) in this thin film regime [6]. The remarkable hydration lubrication properties are often observed in biological systems such as the synovial liquid that lubricates joints, as shown in **Figure 10**, which sparks the study of applications in biomedicine (such as artificial joint) as well as underwater equipment (such as water-lubricated bearings), because the composition of body fluids in living organisms is basically similar to that of seawater. We believe that the research of hydration lubrication could convey extraordinary results in the near future.

6. Conclusions

Electric double layer and hydration effect are crucial for liquid superlubricity, where ion adsorption dominates in hydration superlubricity. As one of IUPAC's 2024 TOP 10 emerging technologies in chemistry, the hydration lubrication provides a framework for understanding, controlling, and designing efficient liquid superlubricity systems in aqueous and biological media at boundary and mixed lubrication regimes. Both strongly hydrated cations and anions, as well as both monovalent and multivalent ions, can achieve hydration superlubricity. For hydrated ions, the stronger the hydration strength and the larger the adsorption density, the higher the carrying capacity and the better the hydration superlubricity performance. Fundamentally, the ion surface coverage dictates the hydration layer structure and its lubricating properties. Based on this mechanism, hydration lubrication can be enhanced through increasing the surface charge density of friction pairs. Therefore, the idea of designing negatively charged modified materials emerges as the times require. New polymer materials grafted by negative charges including modified PEEK and modified UHMWPE exhibit super-low friction and super-low wear in aqueous environment, which has great potential to improve the lubrication performance for water lubricated friction pairs. In addition to polymers which can be modified through graft polymerization, more efforts should be put to obtain hydration lubrication for metals and alloys with extremely low surface charge density.

Acknowledgements

The authors acknowledge the National Natural Science Foundation of China (Grant Nos. 52305198, 51925506, and 52335004), XPLORER PRIZE, China Postdoctoral Science Foundation (No. 2023 M731936), National Excellent Engineers Innovation Institute of Guangdong-Hong Kong-Macao Greater Bay Area (Foshan) Advanced Manufacturing Industry fund (JBGS2024004) and State Key Laboratory of Tribology in Advanced Equipment fund (No. SKLT2024Z05).

Conflict of interest

The authors declare no conflict of interest.

Author details

Tianyi Han*, Chenhui Zhang* and Jianbin Luo*
State Key Laboratory of Tribology in Advanced Equipment, Tsinghua University, Beijing, China

*Address all correspondence to: hanty@mail.tsinghua.edu.cn; chzhang@tsinghua.edu.cn; luojb@tsinghua.edu.cn

References

[1] Israelachvili JN. Intermolecular and Surface Forces. 3rd ed. San Diego, CA: Academic Press; 2011

[2] Zangi R, Engberts JBFN. Physisorption of hydroxide ions from aqueous solution to a hydrophobic surface. Journal of the American Chemical Society. 2005;**127**:2272-2276

[3] Han T, Zhao M, Sun C, Zhao R, Xu W, Zhang S, et al. Macroscale superlubricity of hydrated anions in the boundary lubrication regime. ACS Applied Materials & Interfaces. 2023;**15**: 42094-42103

[4] Gonella G, Backus EHG, Nagata Y, Bonthuis DJ, Loche P, Schlaich A, et al. Water at charged interfaces. Nature Reviews Chemistry. 2021;**5**:466-485

[5] Dreier LB, Nagata Y, Lutz H, Gonella G, Hunger J, Backus EHG, et al. Saturation of charge-induced water alignment at model membrane surfaces. Science Advances. 2018;**4**:eaap7415

[6] Han T, Cao W, Xu Z, Adibnia V, Olgiati M, Valtiner M, et al. Hydration layer structure modulates superlubrication by trivalent La^{3+} electrolytes. Science Advances. 2023;**9**: eadf3902

[7] Martin-Jimenez D, Chacon E, Tarazona P, Garcia R. Atomically resolved three-dimensional structures of electrolyte aqueous solutions near a solid surface. Nature Communications. 2016; **7**:12164

[8] Umeda K, Zivanovic L, Kobayashi K, Ritala J, Kominami H, Spijker P, et al. Atomic-resolution three-dimensional hydration structures on a heterogeneously charged surface. Nature Communications. 2017;**8**:2111

[9] Ma P, Liu Y, Sang X, Tan J, Ye S, Ma L, et al. Homogeneous interfacial water structure favors realizing a low-friction coefficient state. Journal of Colloid and Interface Science. 2022;**626**: 324-333

[10] Li Y, Li S, Bai P, Jia W, Xu Q, Meng Y, et al. Surface wettability effect on aqueous lubrication: Van der Waals and hydration force competition induced adhesive friction. Journal of Colloid and Interface Science. 2021;**599**:667-675

[11] Israelachvili J, Wennerström H. Role of hydration and water structure in biological and colloidal interactions. Nature. 1996;**379**:219-225

[12] Wen S, Huang P. Principles of Tribology. Beijing: Tsinghua University Press; 2012

[13] Luo J, Wen S, Huang P. Thin film lubrication. Part I. Study on the transition between EHL and thin film lubrication using a relative optical interference intensity technique. Wear. 1996;**194**:107-115

[14] Ma L, Luo J. Thin film lubrication in the past 20 years. Friction. 2016;**4**: 280-302

[15] Han T, Zhang S, Zhang C. Unlocking the secrets behind liquid superlubricity: A state-of-the-art review on phenomena and mechanisms. Friction. 2022;**10**: 1137-1165

[16] Raviv U, Klein J. Fluidity of bound hydration layers. Science. 2002;**297**: 1540-1543

[17] Ma L, Gaisinskaya-Kipnis A, Kampf N, Klein J. Origins of hydration lubrication. Nature Communications. 2015;**6**:6060

[18] Diao Y, Espinosa-Marzal RM. The role of water in fault lubrication. Nature Communications. 2018;**9**:2309

[19] Härtl A, Garrido JA, Nowy S, Zimmermann R, Werner C, Horinek D, et al. The ion sensitivity of surface conductive single crystalline diamond. Journal of the American Chemical Society. 2007;**129**:1287-1292

[20] Han T, Xu W, Han J, Adibnia V, He H, Zhang C, et al. Counterion distribution in the stern layer on charged surfaces. Nano Letters. 2024;**24**: 10443-10450

[21] Luxbacher T. The ZETA Guide: Principles of the Streaming Potential Technique. Austria: Anton Paar GmbH; 2014

[22] Brown MA, Abbas Z, Kleibert A, Green RG, Goel A, May S, et al. Determination of surface potential and electrical double-layer structure at the aqueous electrolyte-nanoparticle Interface. Physical Review X. 2016;**6**: 011007

[23] Zhao R, Han T, Zhang C, Yu Q. Disparate external electric field effect on the adsorption and shear behavior of monovalent and trivalent ions in electrolyte solution. Nano Letters. 2024; **24**:7941-7947

[24] Wang R, Klein ML, Carnevale V, Borguet E. Investigations of water/oxide interfaces by molecular dynamics simulations. Wiley Interdisciplinary Reviews-Computational Molecular Science. 2021;**11**:e1537

[25] Leng Y. Hydration force between mica surfaces in aqueous KCl electrolyte solution. Langmuir. 2012;**28**:5339-5349

[26] Funari R, Matsumoto A, de Bruyn JR, Shen AQ. Rheology of the electric double layer in electrolyte solutions. Analytical Chemistry. 2020; **92**:8244-8253

[27] Smith AM, Lee AA, Perkin S. The electrostatic screening length in concentrated electrolytes increases with concentration. The Journal of Physical Chemistry Letters. 2016;7:2157-2163

[28] Elliott GR, Gregory KP, Robertson H, Craig VSJ, Webber GB, Wanless EJ, et al. The known-unknowns of anomalous underscreening in concentrated electrolytes. Chemical Physics Letters. 2024;**843**:141190

[29] Strelcov E, Arble C, Guo H, Hoskins BD, Yulaev A, Vlassiouk IV, et al. Nanoscale mapping of the double layer potential at the graphene–electrolyte Interface. Nano Letters. 2020; **20**:1336-1344

[30] Bañuelos JL, Borguet E, Brown GE Jr, Cygan RT, DeYoreo JJ, Dove PM, et al. Oxide– and silicate–water interfaces and their roles in technology and the environment. Chemical Reviews. 2023;**123**:6413-6544

[31] Li X, Wang ZL, Wei D. Scavenging energy and information through dynamically regulating the electrical double layer. Advanced Functional Materials. 2024;**34**:2405520

[32] Trewby W, Tavakol M, Jaques YM, Voïtchovsky K. Towards local tracking of solvated metal ions at solid-liquid interfaces. Materials Today Physics. 2024;**44**:101441

[33] Dadashi S, Parshotam S, Mandal B, Rehl B, Gibbs JM, Borguet E. Influence of charged site density on local electric fields and polar solvent organization at oxide interfaces. The Journal of Physical Chemistry C. 2024;**128**:9683-9692

[34] Marcelja S. Hydration in electrical double layers. Nature. 1997;**385**:689-690

[35] Zhang C. Hydration force. In: Encyclopedia of Tribology. New York, NY: Springer; 2013. pp. 1704-1708

[36] Dedic J, Okur HI, Roke S. Hyaluronan orders water molecules in its nanoscale extended hydration shells. Science Advances. 2021;7:eabf2558

[37] Gaisinskayakipnis A, Ma L, Kampf N, Klein J. Frictional dissipation pathways mediated by hydrated alkali metal ions. Langmuir. 2016;**32**:4755-4764

[38] Gomollón-Bel F. IUPAC's 2024 top ten emerging technologies in chemistry. Chemistry International. 2024;**46**:8-16

[39] Zhang H, Hou J, Hu Y, Wang P, Ou R, Jiang L, et al. Ultrafast selective transport of alkali metal ions in metal organic frameworks with subnanometer pores. Science Advances. 2018;**4**: eaaq0066

[40] Luo J, Zhou X. Superlubricitive engineering—Future industry nearly getting rid of wear and frictional energy consumption. Friction. 2020;**8**:643-665

[41] Baykara MZ, Vazirisereshk MR, Martini A. Emerging superlubricity: A review of the state of the art and perspectives on future research. Applied Physics Reviews. 2018;**5**:041102

[42] Zhai W, Zhou K. Nanomaterials in superlubricity. Advanced Functional Materials. 2019;**29**:1806395

[43] Ayyagari A, Alam KI, Berman D, Erdemir A. Progress in superlubricity across different media and material systems—A review. Frontiers in Mechanical Engineering. 2022;**8**:908497

[44] Dhanola A, Khanna N, Gajrani KK. A critical review on liquid superlubricitive technology for attaining ultra-low friction. Renewable and Sustainable Energy Reviews. 2022;**165**: 112626

[45] Erdemir A, Martin JM, Jianbin L. Superlubricity, 2nd edition. In: Erdemir A, Martin JM, Luo J, editors. Superlubricity. 2nd ed. Place Elsevier: Elsevier; 2021. pp. i-iii

[46] Yuan S, Chen X, Zhang C. Reducing friction by control of isoelectric point: A potential method to design artificial cartilage. Advanced Materials Interfaces. 2020;7:2000485

[47] Yuan S, Zhang C. A sulfonated modification of PEEK for ultralow friction. Friction. 2022;**10**:1-13

[48] Xu J, Zhang C, Luo J. Hydration lubrication applicable to artificial joints through polyelectrolyte-embedded modification on UHMWPE. ACS Applied Polymer Materials. 2022;**4**(10): 7487-7497

[49] Xu J, Han T, Zhang C, Luo J. Superlubricating UHMWPE composites with functionalized carbon nanotubes reinforcement applicable to artificial joints. Tribology International. 2024;**191**: 109142

[50] Yi S, Guo Y, Li J, Zhang Y, Zhou A, Luo J. Two-dimensional molybdenum carbide (MXene) as an efficient nanoadditive for achieving superlubricity under ultrahigh pressure. Friction. 2023;**11**:369-382

[51] Sun S, Yi S, Li J, Ding Z, Song W, Luo J. Lithium citrate triggered macroscopic superlubricity with near-zero wear on an amorphous carbon film. ACS Applied Materials & Interfaces. 2023;**15**:19705-19714

[52] Han T, Zhang C, Chen X, Li J, Wang W, Luo J. Contribution of a tribo-induced silica layer to macroscale

superlubricity of hydrated ions. The Journal of Physical Chemistry C. 2019; **123**:20270-20277

[53] Ge X, Li J, Wang H, Zhang C, Liu Y, Luo J. Macroscale superlubricity under extreme pressure enabled by the combination of graphene-oxide nanosheets with ionic liquid. Carbon. 2019;**151**:76-83

[54] Singh S, Han T, Chen X, Zhang C. Fabrication and assessment of dry sliding behavior of Ti3C2Tx-MXene reinforced nickel aluminide composites. Tribology International. 2024;**200**:110131

[55] Yang Y, Chen Y, Han T, Shang L, Liu B, Pan M, et al. Evolution of interfacial nanostructures with temperature governing fretting wear in diamond-like carbon films. Carbon. 2024;**226**:119217

[56] Zhang Z, Shen C, Zhang P, Xu S, Kong L, Liang X, et al. Fundamental, mechanism and development of hydration lubrication: From bio-inspiration to artificial manufacturing. Advances in Colloid and Interface Science. 2024;**327**:103145

[57] Israelachvili JN, Adams GE. Measurement of forces between two mica surfaces in aqueous electrolyte solutions in the range 0–1000 nm. Journal of the Chemical Society Faraday Transactions Physical Chemistry in Condensed Phases. 1978;**74**:975-1001

[58] Pashley RM. DLVO and hydration forces between mica surfaces in Li^+, Na^+, K^+, and Cs^+ electrolyte solutions: A correlation of double-layer and hydration forces with surface cation exchange properties. Journal of Colloid & Interface Science. 1981;**83**:531-546

[59] Pashley RM. Hydration forces between mica surfaces in aqueous electrolyte solutions. Journal of Colloid and Interface Science. 1981;**80**:153-162

[60] Israelachvili JN, Pashley RM. Molecular layering of water at surface and origin of repulsive hydration forces. Nature. 1983;**306**:249-250

[61] Ducker WA, Senden TJ, Pashley RM. Direct measurement of colloidal forces using an atomic force microscope. Nature. 1991;**353**:239-241

[62] Ducker WA, Senden TJ, Pashley RM. Measurement of forces in liquids using a force microscope. Langmuir. 1992;**8**: 1831-1836

[63] Donose BC, Vakarelski IU, Higashitani K. Silica surfaces lubrication by hydrated cations adsorption from electrolyte solutions. Langmuir. 2005;**21**: 1834-1839

[64] Taran E, Donose BC, Vakarelski IU, Higashitani K. pH dependence of friction forces between silica surfaces in solutions. Journal of Colloid and Interface Science. 2006;**297**:199-203

[65] Lin W, Klein J. Control of surface forces through hydrated boundary layers. Current Opinion in Colloid & Interface Science. 2019;**44**:94-106

[66] Umeda K, Kobayashi K, Minato T, Yamada H. Atomic-level viscosity distribution in the hydration layer. Physical Review Letters. 2019;**122**: 116001

[67] Zhang S, Zhang C, Hu Y, Ma L. Numerical simulation of mixed lubrication considering surface forces. Tribology International. 2019;**140**: 105878

[68] Zhang S, Zhang C. A new deterministic model for mixed lubricated point contact with high accuracy. Journal of Tribology. 2021;**143**:102201

[69] Zhang S, Zhao M, Zhang C. Plastic–elastic model for water-based lubrication considering surface force. Chinese Journal of Mechanical Engineering. 2022; **35**:117

[70] Butt H-J, Graf K, Kappl M. Physics and Chemistry of Interfaces. 4th ed. Weinheim: Wiley-VCH; 2023

[71] Han T, Zhang C, Luo J. Macroscale superlubricity enabled by hydrated alkali metal ions. Langmuir. 2018;**34**: 11281-11291

[72] Han T, Zhang C, Li J, Yuan S, Chen X, Zhang J, et al. Origins of superlubricity promoted by hydrated multivalent ions. The Journal of Physical Chemistry Letters. 2020;**11**:184-190

[73] Pashley RM. Forces between mica surfaces in La^{3+} and Cr^{3+} electrolyte solutions. Journal of Colloid & Interface Science. 1984;**102**:23-35

[74] Pashley RM, Israelachvili JN. DLVO and hydration forces between mica surfaces in Mg^{2+} , Ca^{2+} , Sr^{2+} , and Ba^{2+} chloride solutions. Journal of Colloid & Interface Science. 1984;**97**:446-455

[75] Pashley RM, Quirk JP. The effect of cation valency on DLVO and hydration forces between macroscopic sheets of muscovite mica in relation to clay swelling. Colloids and Surfaces. 1984;**9**:1-17

[76] Kjellander R, Marčelja S, Pashley RM, Quirk JP. A theoretical and experimental study of forces between charged mica surfaces in aqueous $CaCl_2$ solutions. Journal of Chemical Physics. 1990;**92**:4399-4407

[77] Kohonen MM, Karaman ME, Pashley RM. Debye length in multivalent electrolyte solutions. Langmuir. 2000;**16**: 5749-5753

[78] Perkin S, Goldberg R, Chai L, Kampf N, Klein J. Dynamic properties of confined hydration layers. Faraday Discussions. 2009;**141**:399-413

[79] Tan Q, Zhao G, Qiu Y, Kan Y, Ni Z, Chen Y. Experimental observation of the ion–ion correlation effects on charge inversion and strong adhesion between mica surfaces in aqueous electrolyte solutions. Langmuir. 2014;**30**: 10845-10854

[80] Valmacco V, Elzbieciak-Wodka M, Herman D, Trefalt G, Maroni P, Borkovec M. Forces between silica particles in the presence of multivalent cations. Journal of Colloid and Interface Science. 2016;**472**:108-115

[81] Li J, Zhang C, Deng M, Luo J. Investigation of the difference in liquid superlubricity between water- and oil-based lubricants. RSC Advances. 2015;**5**: 63827-63833

[82] Zhang C, Li K, Luo J. Superlubricity with nonaqueous liquid. In: Erdemir A, Martin JM, Luo J, editors. Superlubricity. 2nd ed. Place Elsevier: Elsevier; 2021. pp. 379-403

[83] Han T, Yi S, Zhang C, Li J, Chen X, Luo J, et al. Superlubrication obtained with mixtures of hydrated ions and polyethylene glycol solutions in the mixed and hydrodynamic lubrication regimes. Journal of Colloid and Interface Science. 2020;**579**:479-488

[84] Li S, Bai P, Li Y, Jia W, Li X, Meng Y, et al. Extreme-pressure superlubricity of polymer solution enhanced with hydrated salt ions. Langmuir. 2020;**36**: 6765-6774

[85] Lin W, Klein J. Recent progress in cartilage lubrication. Advanced Materials. 2021;**33**:2005513

[86] Rosenhek-Goldian I, Kampf N, Klein J. Trapped aqueous films lubricate

highly hydrophobic surfaces. ACS Nano. 2018;**12**:10075-10083

[87] Li Z, Liu Q, Zhang D, Wang Y, Zhang Y, Li Q, et al. Probing the hydration friction of ionic interfaces at the atomic scale. Nanoscale Horizons. 2022;**7**:368-375

[88] Dong C, Liu Y, Han T, Zhou X, Meng Y, Tian Y, et al. Effect of hydrated ions and wettability on ice friction. Friction. 2024. DOI: 10.26599/FRICT.2025.9440972

[89] Dong C, Liu Y, Meng Y, Du S, Zhu S, Tian Y, et al. Ion-specific ice provides a facile approach for reducing ice friction. Journal of Colloid and Interface Science. 2024;**675**:451-460

[90] Li Z, Liu Q, Li Q, Dong M. The role of hydrated anions in hydration lubrication. Nano Research. 2023;**16**: 1096-1100

[91] Wang Z, Guo H, Singh S, Adibnia V, He H, Kang F, et al. Electric potential controlled ionic lubrication. Lubricants. 2024;**12**:214

[92] Li S, Liu C, He W, Zhang J, Qiao X, Li J, et al. A review of electric potential-controlled boundary lubrication. Lubricants. 2023;**11**:467

[93] Spikes HA. Triboelectrochemistry: Influence of applied electrical potentials on friction and wear of lubricated contacts. Tribology Letters. 2020;**68**:90

[94] Xie G, Guo D, Luo J. Lubrication under charged conditions. Tribology International. 2015;**84**:22-35

[95] Bresme F, Kornyshev AA, Perkin S, Urbakh M. Electrotunable friction with ionic liquid lubricants. Nature Materials. 2022;**21**:848-858

[96] Meng Y, Liu C. Spatiotemporal manipulation of boundary lubrication by electro-charging and electrochemical methods. In: Erdemir A, Martin JM, Luo J, editors. Superlubricity. 2nd ed. Place Elsevier: Elsevier; 2021. pp. 499-516

[97] He S, Meng Y, Tian Y. Correlation between adsorption/desorption of surfactant and change in friction of stainless steel in aqueous solutions under different electrode potentials. Tribology Letters. 2011;**41**:485-494

[98] Yang X, Meng Y, Tian Y. Potential-controlled boundary lubrication of stainless steels in non-aqueous sodium dodecyl sulfate solutions. Tribology Letters. 2014;**53**:17-26

[99] Liu C, Fang J, Wen X, Tian Y, Meng Y. Active control of boundary lubrication of ceramic Tribo-pairs in sodium dodecyl sulfate aqueous solutions. Tribology Letters. 2021;**69**:144

[100] Liu C, Li X, Li X, Li W, Tian Y, Meng Y. On-line feedback control of sliding friction of metals lubricated by adsorbed boundary SDS films. Lubricants. 2022;**10**:148

[101] Zhou X, Xiang Z, Bai P, Wen X, Dong C, Liu C, et al. Voltage-controlled finger friction behavior for immersive tactile experience in aqueous environment. Tribology International. 2024;**194**:109487

[102] Tivony R, Zhang Y, Klein J. Modulating interfacial energy dissipation via potential-controlled ion trapping. The Journal of Physical Chemistry C. 2021;**125**:3616-3622

[103] Li S, Bai P, Li Y, Pesika NS, Meng Y, Ma L, et al. Quantification/mechanism of interfacial interaction modulated by electric potential in aqueous salt solution. Friction. 2020;**9**:1-11

[104] Gao T, Li J, Yi S, Luo J. Potential-dependent friction on a graphitic surface in ionic solution. The Journal of Physical Chemistry C. 2020;**124**:23745-23751

[105] Zhang Y, Li H, Wang J, Silvester DS, Warr GG, Atkin R. Potential-dependent superlubricity of stainless steel and Au(111) using a water-in-surface-active ionic liquid mixture. Journal of Colloid and Interface Science. 2025;**678**:355-364

[106] Besteman K, Zevenbergen MA, Heering HA, Lemay SG. Direct observation of charge inversion by multivalent ions as a universal electrostatic phenomenon. Physical Review Letters. 2004;**93**:170802

[107] Li SX, Guan W, Weiner B, Reed MA. Direct observation of charge inversion in divalent nanofluidic devices. Nano Letters. 2015;**15**:5046-5051

[108] Van der Heyden FH, Stein D, Besteman K, Lemay SG, Dekker C. Charge inversion at high ionic strength studied by streaming currents. Physical Review Letters. 2006;**96**:224502

[109] Besteman K, Zevenbergen M, Lemay S. Charge inversion by multivalent ions: Dependence on dielectric constant and surface-charge density. Physical Review E. 2005;**72**: 061501

[110] Li H, Xu Z, Ma M. Temperature-dependent slip length for water and electrolyte solution. Journal of Colloid and Interface Science. 2023;**636**:512-517

[111] Xu Z, Li H, Ma M. Accurate estimation of dynamical quantities for nonequilibrium nanoscale systems. Physical Review E. 2023;**107**:014124

[112] Qiu Y, Ma J, Chen Y. Ionic behavior in highly concentrated aqueous solutions nanoconfined between discretely charged silicon surfaces. Langmuir. 2016;**32**:4806-4814

[113] Lin W, Kluzek M, Iuster N, Shimoni E, Kampf N, Goldberg R, et al. Cartilage-inspired, lipid-based boundary-lubricated hydrogels. Science. 2020;**370**:335-338

[114] Hu T, Zhang W, Wu M, Tang W, Chen X, Li X, et al. Ultralow friction of PEEK composites under seawater lubrication. Journal of Materials Research and Technology. 2024;**30**: 983-991

[115] Wu M, Hu T, Lan J, Li X, Zhang C, Luo J. Carbon fibre reinforced bisphenol-a type SPEEK composite for water lubricated superlubricity. Composites Part A: Applied Science and Manufacturing. 2024;**180**:108113

[116] Xu J, Han T, Zhang C, Luo J. Differential hydration lubrication performance of polyelectrolyte-modified UHMWPE promoted by diverse charge characteristics. Friction. 2024. DOI: 10.26599/FRICT.2025.9441006

[117] Lan J, Mo C, Chen X, Hu T, Li X, Zhang C. Carbon fiber/nano SiO_2 reinforced polyelectrolyte-graft UHMWPE for water lubricated superlubricity. Tribology International. 2025;**202**:110303

[118] Moro T, Takatori Y, Ishihara K, Konno T, Takigawa Y, Matsushita T, et al. Surface grafting of artificial joints with a biocompatible polymer for preventing periprosthetic osteolysis. Nature Materials. 2004;**3**:829-836

[119] Klein J. Hydration lubrication. Friction. 2013;**1**:1-23

Chapter 5

The Wettability of PMMA Surface by Aqueous Solutions of Anionic Gemini Surfactants with Different Hydrophobic Carbon Chain Lengths

Zhe Hang, Lu Zhang, Fenrong Liu and Wangjing Ma

Abstract

This study investigates the synthesis of several new Gemini surfactants with various hydrophobic carbon chain length amide groups (C12, C14, C16, and C18) using triethylenetetramine, maleic anhydride, and internal olefins with different carbon chain lengths as raw materials. The research examined the adsorption mechanism and hydrophilic modification capacity of four different kinds of Gemini surfactants on the polymethyl methacrylate (PMMA) surface. Surface tension and contact angle data for each Gemini surfactant were used to compute the adhesion work and adhesion tension. According to the surface activity parameters, the critical micelle concentration and maximum adsorption amount decrease with the increase in the length of the hydrophobic carbon chain. However, the equilibrium surface tension first decreases and then increases with the increase in the length of the hydrophobic carbon chain. The four surfactants exhibit complex self-aggregation behavior in the solution due to their long hydrophobic chain structure and flexible spacer. The four surfactants are heavily adsorbed on the PMMA surface, forming semicolloidal aggregates, according to the combination of contact angle measurements, adhesion tension, and interfacial tension data. Moreover, compared with literature, the four surfactants synthesized in this study show better hydrophilic modification ability on the PMMA surface.

Keywords: Gemini surfactant, PMMA, hydrophilic modification, wetting effect, adsorption mechanism

1. Introduction

In the study of solid-liquid interfaces, wetting behavior is one of the key characteristics, and surfactants are indispensable. Surfactants play a crucial role in many applications, including enhanced oil recovery, emulsification, and wettability, where wetting phenomena are important [1–4]. Scholars have been attempting to alter the chemical structure of surfactants to improve their wetting performance. As surfactants have developed over time, they have gone from being single-chain to having a variety of unique structures, such as zwitterionic surfactant [5, 6], double-chained surfactants with dual hydrophobic chains [7], inserted functionalized monomeric units [8], amphoteric ions [9], and all of the above. Gemini and double-chained surfactants are two of them that have attracted a lot of attention lately. Especially, on the surface of polymethyl methacrylate (PMMA), notable advancements have been observed in the adsorption behavior and wetting characteristics of Gemini surfactants and double-chain surfactants, indicating their merit for more study.

The properties of PMMA include strong dimensional stability, weather resistance, and flexibility. It has many uses nowadays because of its superior chemical and physical characteristics [10–13]. Many functional groups, including -CH3, -CO, and -OCH3 [14], are present on the solid surface of PMMA. These functional groups cause the adsorption behavior of surfactants on the PMMA surface to become complicated. As a result, researching the wettability of PMMA surfaces is essential.

According to recent study, using Gemini surfactants may boost oil recovery rates and enhance oil displacement efficiency [14–16]. Gemini surfactants also have antioxidant and chemical stability, which makes it possible for them to successfully shield metal surfaces from aging and chemical corrosion [17–19]. Gemini surfactants have a lower critical micelle concentration (CMC) and a higher surface activity than traditional surfactants. There is, however, little information on the adsorption behavior of anionic Gemini surfactants with different hydrophobic carbon chain lengths on PMMA surfaces, in contrast to reports on the adsorption behavior of zwitterionic surfactants [20], cationic Gemini surfactants [21], and anionic Gemini surfactants [22] on PMMA surfaces.

In recent years, many research on the adsorption processes and wetting properties of different kinds of surfactants on PMMA surfaces have been published. Anionic and cationic surfactant combinations were studied by Wang et al. [21] to improve the wettability of weakly polar PMMA and nonpolar PTFE. The blend of anionic and cationic surfactants adsorbed on the PTFE surface results in a significant reduction in the contact angle, suggesting enhanced wettability. On the other hand, the anionic and cationic surfactant mixture's initial adsorption on the PMMA surface increases its hydrophobicity, while further adsorption increases its hydrophilicity. Moreover, the adsorption capacity of the anionic and cationic surfactant mixture on the PTFE surface is stronger than on the PMMA surface, and the adsorption on the PTFE surface results in a more pronounced reduction in the contact angle. Zhang et al. [23] investigated the wetting modification and adsorption behavior of single-chain alkyl C3 and C6 monomers as well as Gemini anionic surfactants on PMMA surfaces. Single-chain structured surfactants form hemimicelles on the PMMA surface, whereas Gemini surfactants adsorb onto the PMMA surface through a bilayer structure, according to calculations and measurements of contact angles and surface tensions. However, the adsorption of Gemini C3 and C6 at the air-liquid interface varies due to different spacer lengths and dual hydrophobic chains. Owing to its more compact structure, C3 causes self-aggregation in solution, which produces two critical micelle concentrations (CMCs) and two plateau values for surface tension. Conversely, C6 has a single plateau value, just like other surfactants. Du et al. [24] and colleagues also looked at some surfactants that have special wetting qualities because they have branched hydrophobic chains or polyoxyethylene (PO) units in their structure; adding PO groups lowers the surfactant's CMC considerably, and the CMC goes down as the number of PO groups increases. Below the CMC, the γ_{SL} value increases linearly, and above the CMC, it decreases linearly. The addition of PO groups increases the surfactant's hydrophobic and hydrophilic modification capabilities.

We have discovered that double-chain surfactants and Gemini surfactants offer a wide range of potential applications based on the previously mentioned research. Thus, four different types of alkyl maleic acid surfactants with varying hydrophobic chain lengths are synthesized in this research. By examining the adsorption processes of Gemini surfactants with varying hydrophobic carbon chain lengths, we were able to better understand the adsorption behavior and wetting qualities on PMMA surfaces. This work is to clarify the action mechanism of adsorption on the polymer surface, unveil the adsorption behavior on the polymer surface, identify the rules governing changes in adsorption equilibrium, and investigate potential applications in the modification of polymer materials. Some of the theoretical and practical implications of this work include understanding the hydrophilic and hydrophobic modification capabilities of Gemini anionic surfactants with varying hydrophobic carbon chain lengths on polymer surfaces and developing solid surface hydrophilic-hydrophobic technologies.

This work methodically examines the Gemini anionic surfactants' wetting characteristics on the surface of PMMA with different hydrophobic carbon chain lengths. Different hydrophobic chain lengths of Gemini anionic surfactants were produced. The droplet method was utilized to measure the contact angle at the gas-liquid-solid interface, whereas the plate method was employed to determine the surface tension. Examine how anionic Gemini surfactants with varying hydrophobic chain lengths wet on the PMMA surface by analyzing the data that was collected.

2. Experimental

2.1 Materials

The patented catalyst [25] from earlier studies was utilized to prepare alkyl succinic anhydrides with varying carbon chain lengths. It has two different length alkyl chains. The produced compounds were then transformed into Gemini surfactants. **Figure 1** depicts the chemical structure of surfactants based on Dix's [26] synthesis. The chemical structure was ascertained by mass spectrometry and proton nuclear magnetic resonance (H-NMR) spectroscopy. In the experiment, ultrapure water with a resistivity of 18.2 MΩ cm was utilized.

2.2 Surface tension measurement

Gemini surfactants were measured for surface tension using the Wilhelmy plate technique. After meticulously cleaning the platinum plate with a tensiometer (Data Physics AG, DCAT 21), it was heated with a flame to 298 ± 0.5 K, which produced a vivid red light. In every case, measurements were taken more than three times, with a maximum standard deviation of 0.2 mN m^{-1}.

2.3 Contact angle measurement

The contact angles of surfactant aqueous solutions on PMMA plates were measured at (298 ± 0.5 K) using the OCA 20 contact angle measurement method from

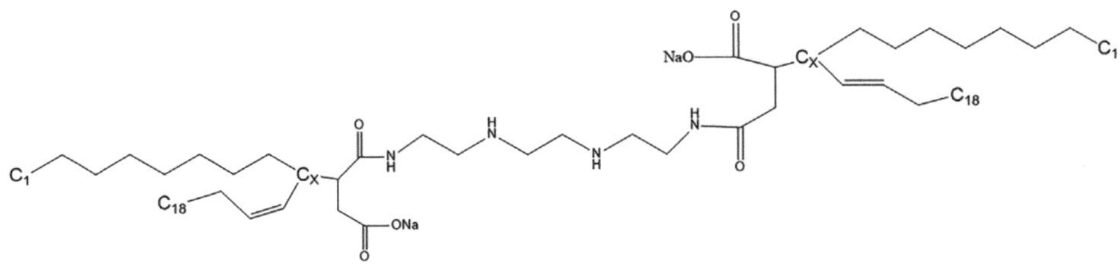

C18, $Na_2(C_{50}H_{92}O_6N_4)$

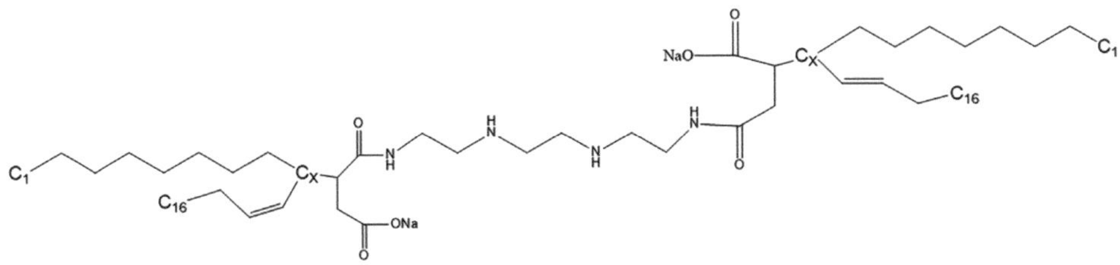

C16, $Na_2(C_{46}H_{84}O_6N_4)$

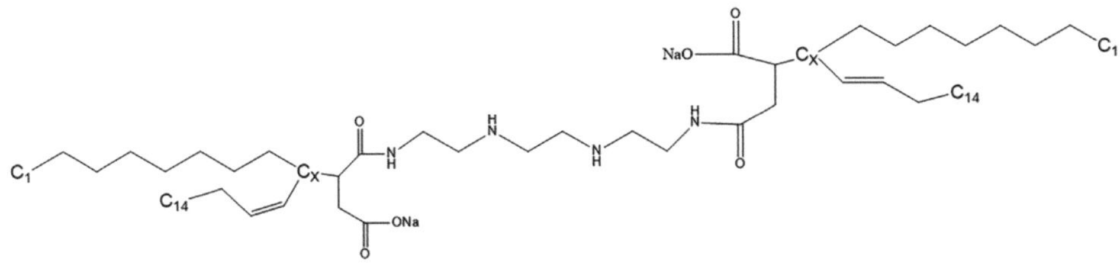

C14, $Na_2(C_{42}H_{76}O_6N_4)$

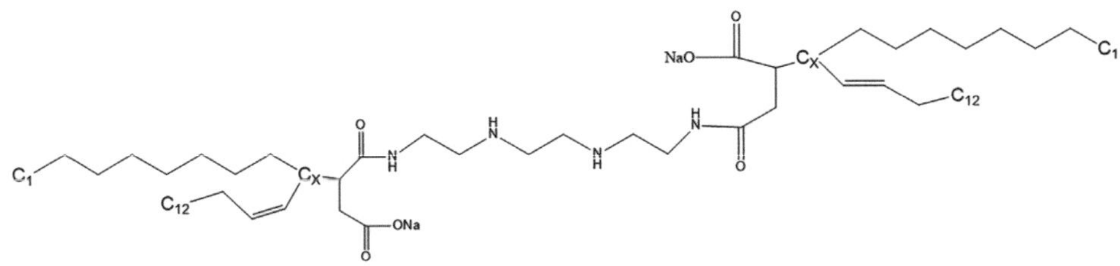

C12, $Na_2(C_{38}H_{68}O_6N_4)$

Figure 1.
Molecular structures and chemical formulas of surfactants C12, C14, C16, and C18: CX represents X carbon atoms at position X, where for C12 ($2 \leq X \leq 11$), for C14 ($2 \leq X \leq 13$), for C16 ($2 \leq X \leq 15$), and for C18 ($2 \leq X \leq 17$).

Data Physics Instruments GmbH, Germany. Repeat measurements were conducted at different spots along the solid surface after the droplet had settled and measurements were obtained from both sides. The standard deviation was less than 3°.

To measure contact angles, PMMA plates were cut from a larger sheet. The plates were cleaned in an ultrasonic bath for half an hour after being rinsed with ultrapure water and ultrasonically cleaned with ethanol. These surfaces were then heated to 373 K for 2 hours.

3. Results and discussion

3.1 Surface tension of surfactants solution

The interfacial adsorption behavior of surfactants is reflected in surface tension. The Gemini surfactants from C12 to C18 have CMC values of 1.49×10^{-4}, 1.57×10^{-4}, 0.68×10^{-4}, and 0.47×10^{-4} mol l^{-1}, respectively, as shown in **Figure 2**. The surface tension values drop quickly as the surfactant concentration rises and each one reaches the CMC plateau.

The Gibbs adsorption Eq. (1) and the graph's slope are used to determine the maximum interface excess concentration.

$$\Gamma = -\left(\frac{1}{2.303nRT}\right)\left(\frac{dy}{dlogC}\right) \tag{1}$$

The surface tension (mN/m), molar concentration (mol/L), gas constant (8.314 J/mol·K), and absolute temperature (T) are represented by the symbols γ, C, and R, respectively. Three ions are ionized among the four Gemini surfactants; n is assumed to be 3. The minimal area Amin at the air-water interface for every molecule may be found using formula (2) as follows: the air-water interface for each molecule:

$$A_{min} = \left(\frac{10^{14}}{N_A \Gamma_{max}}\right) \tag{2}$$

where NA is the Avogadro constant.

The values of the CMC, γ_{CMC}, Γ_{max}, and A_{min} are listed in **Table 1**.

The surface tension slope that was discussed in the first paragraph is the basis for the calculation of the adsorption area and capacity in **Table 1**. **Table 1** shows that as the hydrophobic chain's length increases, the surfactant's CMC progressively drops at the same temperature. This is because the product's hydrophobic impact rises with

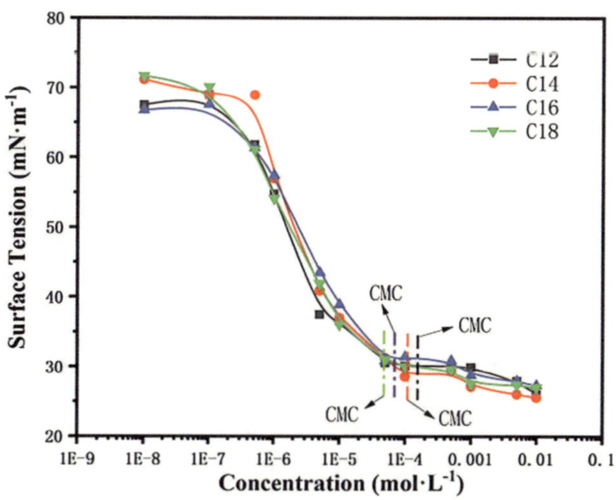

Figure 2.
Surface tension versus of the concentration for C12, C14, C16 and C18.

Surfactant	CMC/(10^{-4} mol L^{-1})	γ_{CMC} (mNm^{-1})	$10^{10}\Gamma_{max}$ (Mol cm^{-2})	A_{min} (nm^2)
C12	1.49	30.17	0.93	1.77
C14	1.11	28.96	1.09	1.52
C16	0.68	31.35	0.92	1.79
C18	0.47	31.52	0.90	1.85

Table 1.
The surface activity properties of C12, C14, C16, and C18.

lengthening of the hydrophobic carbon chain, increasing the likelihood that the surfactant would combine at the contact. As a result, as the number of carbon atoms in the hydrophobic carbon chain increases, the CMC value steadily drops. In addition, the amide bonding partially mitigates the electrostatic repulsion between the two polar head groups, resulting in Gemini surfactants that are closely packed and relatively stable in solution. The lower CMC is caused by Gemini surfactants' propensity to form micelles at low concentrations.

C14 has the greatest impact on lowering water's surface tension, with a value of 28.96 mNm^{-1}. The surface tension first falls and then progressively rises as the hydrophobic chain lengthens from C12 to C18. This is because surfactant molecules arrange themselves at the air/liquid interface in a way that causes chain curling and folding as the hydrophobic carbon chain lengthens. This, in turn, causes the hydrophobic chain arrangement of the surfactant to become less compact. As a result, fewer surfactant molecules with lengthy hydrophobic chains are adsorbed at the air-liquid interface. The C14 surfactant has the greatest capacity to lower surface tension because it is more easily compressed than the C12 surfactant and has the greatest number of molecules arranged at the air-liquid interface.

As the hydrophobic chain's carbon atom count rises, the Γ_{MAX} of products C12–C14 falls while the A_{min} rises. This is specifically because it is difficult for the long hydrophobic carbon chains to align vertically on the air-water interface due to their propensity for bending and folding, which increases the area of a single molecule on the contact [27].

3.2 Contact angles on the PMMA surface

The difference in contact angles between four surfactant solutions on PMMA is seen in **Figure 3**. With no discernible variations, the contact angles of the four surfactants range from 1E-8 mol/L to 1E-4 mol/L by about 76°. On the other hand, the C12 and C14 contact angle dramatically reduces when the concentration rises over 1E-4 mol/L. On the other hand, the concentration of the longer carbon chain surfactants C16 and C18 remains constant until a sudden drop occurs. At a concentration of 0.01 mol/L, the contact angles of surfactants C12, C14, C16, and C18 on the PMMA surface are decreased to 34.3°, 47.3°, 44.4°, and 54.7°, respectively.

Furthermore, it was discovered that these Gemini surfactants' contact angle fluctuation on the PMMA surface differs noticeably from that of conventional surfactants as reported in the literature. Other widely used surfactants, such as sodium dodecyl sulfate (SDS) and triton X-100 (TX-100), demonstrate a reduction in contact angle as concentration increases on the PMMA surface. However, they become stable when the concentration reaches the CMC [28]. For SDS and TX-100, the minimum contact angles are roughly 55° and 35°, respectively. Prior studies have demonstrated that

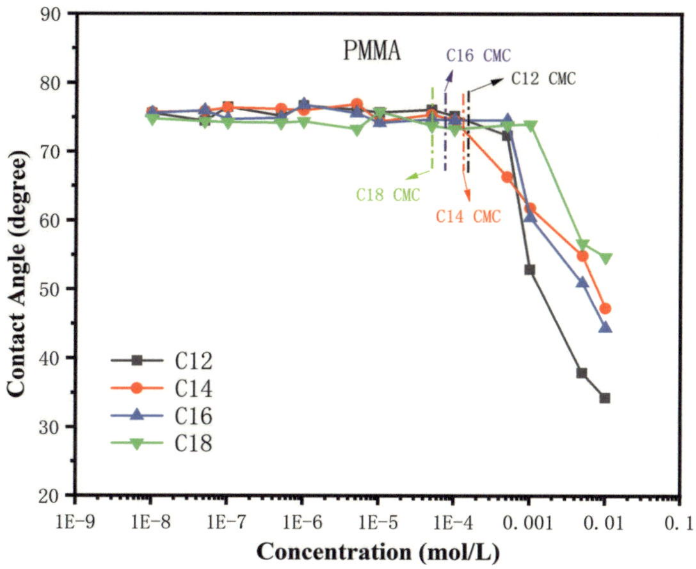

Figure 3.
The contact angle (y) of C12, C14, C16, and C18 on PMMA varies with concentration.

while the hydrophobic chain length gradually increases the surfactant's hydrophobicity at the surface [29], changes in the bridge length of these microsurfactants have no effect on changes in the contact angle on the PMMA surface [23].

3.3 Adhesion tension on the PMMA surface

Adhesion tension, which is determined by the equilibrium contact angle, Young's equation, and three interfacial free energies, is a crucial factor in assessing the wettability of solid surfaces.

$$\gamma_{LV}\mathrm{COS}\gamma = \gamma_{SV} - \gamma_{SL} \tag{3}$$

where the solid-gas, solid-liquid, and gas-liquid interfacial free energies are represented, respectively, by γ_{SV}, γ_{SL}, and γ_{LV}. The link between the adsorption capacities at the gas-liquid, solid-liquid, and gas-solid interfaces can be ascertained using the Young and Gibbs equations.

$$\frac{d(\gamma_{LV}COS\theta)}{d\gamma_{LV}} = \frac{\Gamma_{SV} - \Gamma_{SL}}{\Gamma_{LV}} \tag{4}$$

For surfactants, Γ_{SV}, Γ_{SL}, and Γ_{LV} stand for the maximal surface adsorption capacity at the solid-gas, solid-liquid, and gas-liquid interfaces, respectively. One may determine the value of Γ_{SL}/Γ_{SL} given $\Gamma_{SV} = 0$ by calculating the slope of the γ_{LV} cosθ and Γ_{LV} curves under CMC.

When surfactant molecules' hydrophilic groups adhere to a solid surface, in this instance, the solid-liquid and gas-liquid adsorption capabilities have a negative ratio:

$$\frac{\Gamma_{SL}}{\Gamma_{LV}} = -a \tag{5}$$

Alkyl chains are adsorbed on solid surfaces by hydrophobic interactions. As a result, the ratio of adsorption capacity at solid-liquid interfaces is positive:

$$\frac{\Gamma_{SL}}{\Gamma_{LV}} = a \tag{6}$$

Because hydrophobic tails and PMMA interact less strongly than conventional ionic surfactants do, surfactant molecules adsorb on the surface, lowering the solid-liquid interfacial free energy and producing a negative slope. Surfactants with a positive slope typically contain multiple hydrophilic groups. Based on this study, the surfactant synthesized bears an ionic head containing a -COO- and -CONH- group, which can form hydrogen bonds with PMMA, and tightly adhering to the surface.

Figure 4 illustrates surface tension and adhesion tension in relation to one another for two types of surfactants. In the initial phase, both surfactants exhibit distinct linear relationships between surface tension and adhesion tension. A constant surface tension occurs when the concentration reaches CMC. These four surfactants have slopes of 0.24, 0.22, 0.23, and 0.26. We may calculate that 4.2, 4.5, 4.3, and 3.8 times more adsorption takes place at the air-liquid interface using these four surfactant slopes. A saturated adsorption film forms at the air-liquid interface when surfactant molecules reach saturation, thereby maintaining surface tension. In the meantime, adhesion tension will rise vertically at the solid-liquid interface as surfactant molecules continue to adsorb there.

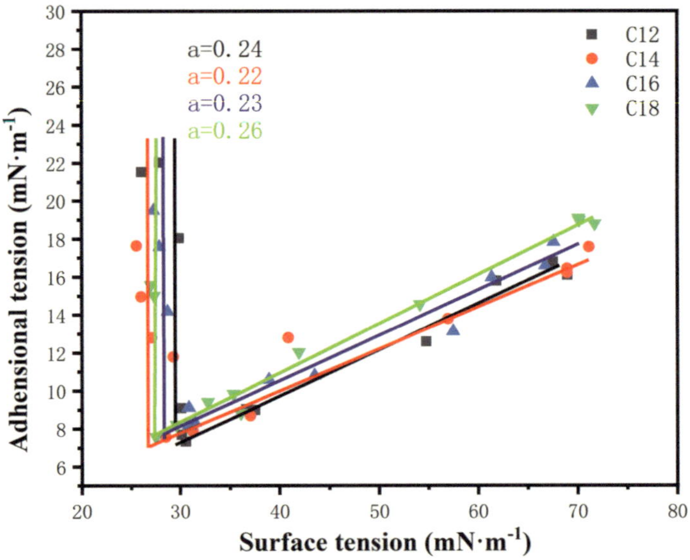

Figure 4.
Adhesion tension of C12, C14, C16, and C18 on the surface of PMMA in relation to surface tension.

3.4 Interfacial tension of the PMMA-solution interfaces

The solid-liquid interfacial tension of these four surfactants was measured to examine their adsorption behavior on the PMMA surface. The PMMA solid has a surface free energy of 39.5 mJ m^2. Eq. (3) can be used to determine the interfacial tension, or γ_{SL}, between PMMA and the liquid.

The interfacial tension γ_{SL} change curve is shown in **Figure 5** along with the surfactant concentrations. Three adsorption phases were noted for each of the four surfactants on PMMA.

At low concentrations, the hydrophobic tails of the four surfactants face the solution, while the hydrophilic parts adsorb onto the PMMA surface. Interfacial tension rises as a result, changing the PMMA surface to a somewhat hydrophobic condition. The four surfactants' interfacial tension reaches a plateau between 1E-5 and 5E-4 mol/l, indicating that they form colloids in the solution and approach saturation in adsorption on monolayers. **Figure 5** illustrates this interfacial tension plateau. When the concentration of the surfactant rises above 5 E-4 mol/L, hydrophobic interactions cause the molecules to adsorb onto a monolayer, forming a hemimicelle, with the hydrophilic parts facing the solution, leading to a gradual decrease in interfacial tension, making the PMMA surface hydrophilic. From **Figure 5**, the low concentration slopes of C12, C14, C16, and C18 are 3.4, 3.65, 4.1, and 4.9, respectively, as can be shows. At high concentrations, the slopes of C12, C14, C16, and C18 are 8.7, 5.17, 7.83, and 8.47 respectively, which are greater than the slopes before the CMC.

Substances that adsorb on the PMMA surface following the CMC do so in three primary ways: (1) monolayer adsorption (via hydrophobic tails or polar groups); (2) bilayer adsorption (where a second layer of surfactants adsorbs on top of the first layer); and (3) aggregation formation, in which hydrophobic interactions cause surfactant molecules to form hemimicelles on a solid surface [24]. The surface of

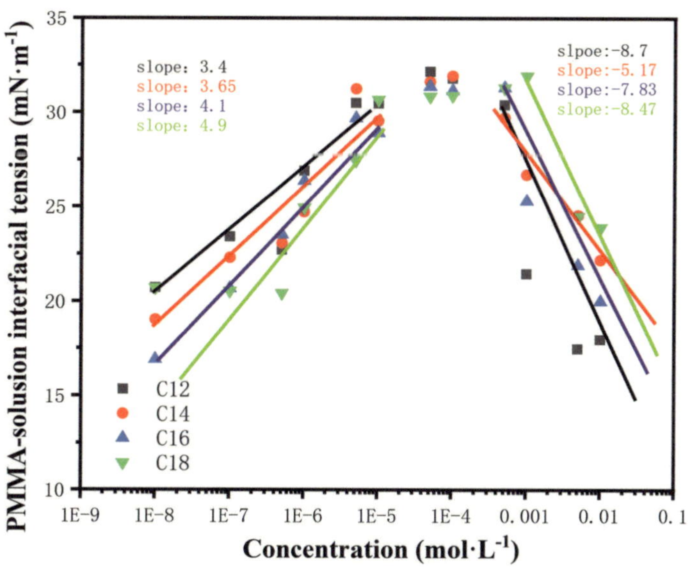

Figure 5.
The interfacial tension of C12, C14, C16, and C18 on the surface of PMMA varies with concentration.

PMMA is continually adsorbing four surfactants, which raises the interfacial tension values. The interfacial tension shows a plateau value in the concentration range of 1E-5 to 5E-4 mol/L as the surfactant molecules' adsorption on the PMMA surface progressively achieves saturation, generating the first layer of saturated adsorption film. The adsorption rate also decreases during this process. At this stage, the solution's micelles begin to form. When surfactant molecules continue to adsorb in substantial quantities onto PMMA at concentrations higher than 5E-4 mol/L, hemicellar aggregates are created. Therefore, the interfacial tension drops rapidly as the slope of the curve rises.

$\Delta_{SL(max)}$ is the difference between the maximum γ_{SL} value and the γ_{SL} value without surfactants, and $\gamma_{SL(min)}$ is the lowest γ_{SL} value at high concentrations. In the meantime, the maximum hydrophobic and maximum hydrophilic capacities of the surfactants are denoted by $\Delta_{SL(max)}$ and $\gamma_{SL(min)}$, respectively. **Table 2** lists the $\Delta_{SL(max)}$ and $\gamma_{SL(min)}$ values for a few surfactants. **Table 2** shows that the corresponding $\gamma_{SL(min)}$ values for C12, C14, C16, and C18 are 17.5, 22.2, 20, and 23.9. According to the findings, C12 modifies the PMMA surface's hydrophilicity the most. The computed values from actual data (A_{mina}) match the theoretical values (A_{minT}) obtained from the Gibbs equation using the slope with adhesive tension, as indicated in **Table 2**, confirming the validity of the experimental results.

The maximum and lowest values of γ_{SL}, which indicate the surfactants' maximal hydrophobic and hydrophilic modification capacities on the PMMA surface, may be determined from the interfacial tension curves. A few surfactants with unique structures were chosen specifically for this study's comparison. **Table 3** contains a list of the data.

In comparison to other structures of betaine, the four types of Gemini surfactants produced in this study show better hydrophobic modification capabilities. C12 exhibits the best hydrophobic modification ability among all, having the shortest hydrophobic chain. This is explained by the Cn ionic head's high adsorption onto the PMMA surface. In contrast, at high concentrations, C12 exhibits excellent hydrophilic modification capabilities, which underscores the advantages of Gemini double-chain surfactants in forming hemimicelles on the PMMA surface.

3.5 Work of adhesion on the PMMA surface

Eq. (7) illustrates that the reversible work necessary to isolate a unit area of liquid from a solid surface defines the work of adhesion of a liquid to a solid (W_A).

$$W_A = \Gamma_{SV} + \Gamma_{LV} - \Gamma_{SL} \tag{7}$$

Surfactant	<CMC $10^{10}\Gamma_{MAX}$ (mol cm^{-2})	<CMC A_{mina} (nm^2)	$A_{minLV}/(\Gamma_{SL}/\Gamma_{LV})$ A_{minT} (nm^2)	>CMC $10^{10}\Gamma_{MAX}$ (mol cm^{-2})	>CMC A_{minb} (nm^2)
C12	0.2	8.37	8.34	0.51	3.27
C14	0.21	7.79	7.78	0.30	5.50
C16	0.24	6.94	6.94	0.46	3.63
C18	0.29	5.81	5.86	0.50	3.36

Table 2.
Interfacial parameters of C12, C14, C16 and C18.

Samples	$\Delta_{SL\ (max)}$/ (mN m^{-1})	$\gamma_{SL\ (min)}$ / (mN m^{-1})
C12	14.7	17.5
C14	10	22.2
C16	11.3	20
C18	11.5	23.9
18S [30]	9.2	24
18C [30]	9	21
GeminiC6 [23]	10.37	26.3
GeminiC3 [23]	7.26	24.9
C_{16}GPC [31]	2.7	18.8
$C_{16}G(EO)_3$PC [31]	4.4	18
C_{16}GPB [31]	5	18
$C_{16}G(EO)_3$PB [31]	4.9	17.5
S-$C_{12}PO_7$S [24]	10.5	11.1
S-$C_{12}PO_{13}$S [24]	11.8	10.6

Table 3.
Maximum hydrophobicity of PMMA surface modification capacity ($\Delta SL(max)$) and maximum hydrophilic modification capacity ($\gamma SL(min)$).

WA may also be computed by using the Young formula and Eq. (8).

$$W_A = \Gamma_{LV}(\text{COS}\theta + 1) \tag{8}$$

Eq. (8) may be used to determine the adhesion work based on the surface tension and contact angle values of C12, C14, C16, and C18, as seen in **Figure 6**.

The higher the adhesion work, the more easily the surfactant solution wets the PMMA surface. Eq. (7) states that the values of γ_{SL}, γ_{SV}, and γ_{LV} define the value of the adhesion work. Since γ_{SV} has a constant value, the value of the adhesion work equals the difference between γ_{SL} and γ_{LV}. As the surfactant concentration rises, γ_{SL} increases and γ_{LV} declines below the CMC. Consequently, the value of the adhesion work decreases sharply below the CMC. Following the CMC, when surfactant concentration rises more, γ_{SL} falls while γ_{LV} remains rather constant. As a result, adhesion work becomes more valuable after the CMC. It is evident that C12 is most suited for wetting the PMMA surface since it shows the maximum adhesion work after the CMC.

3.6 Mechanism responsible for surfactant adsorption behavior on the PMMA surface

The solid-liquid interfacial tension, contact angle, and surface tension variations as functions of concentration are shown in the picture to aid in our understanding of the four surfactants' adsorption process on the PMMA surface. In addition, mechanism diagrams were created to clearly show the adsorption behavior of the four surfactants on the PMMA surface using examples of C12 and C18, which contain the shortest and longest hydrophobic carbon chains, respectively. This mechanism

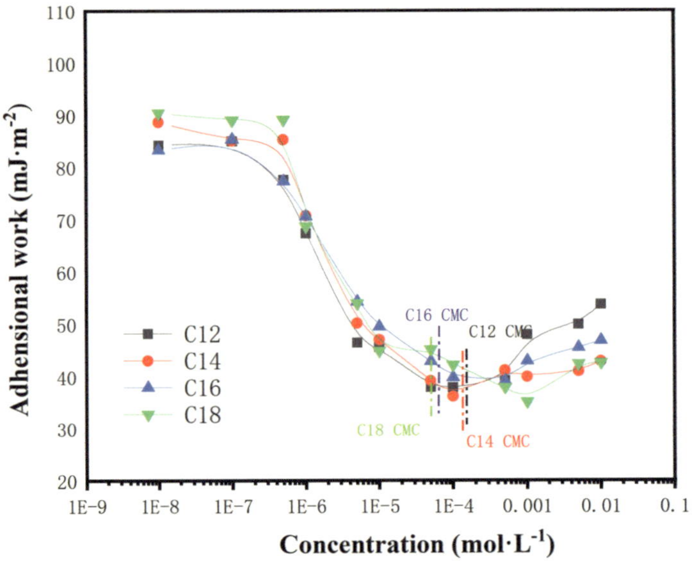

Figure 6.
Effect of GeminiC6, C3 and monomer concentration on the adhesion function WA of the PMMA surface.

diagram is shown in **Figure** 7. Based on the data from **Figures** 7 and **8**, we will discuss C12 and C14 in three stages, while C16 and C18 will be discussed in four stages.

From 1×10^{-7} mol/l to the CMC, the concentration is in the first stage. Surface tension, interfacial tension, and contact angle all hardly alter during this period. This is due to the surfactant's low bulk concentration, which makes it less effective in changing the PMMA surface.

A rise in concentration characterizes the second stage, as surfactant molecules quickly adsorb at the air-liquid interface to form surface micelles and reach the CMC. Surfactant molecules start to adsorb quickly at the interfaces between solids and liquids as well as between air and liquid, as the surfactant concentration progressively rises. As a result, surface tension decreases and interfacial tension rises. Through hydrogen bonding, the ionic heads of the four surfactants adsorb onto the PMMA surface because their hydrophobic alkyl chains face the solvent. Because the drop in surface tension is balanced by the rise in solid-liquid interfacial tension, the contact angle stays constant.

The concentration ranges from the CMC to 1×10^{-2} mol/l is referred to as the third stage. At the gas-liquid interface, molecules of the surface-active agent become saturated and form micelles in the solution. Simultaneously, molecules of the surface-active agent create a saturated layer of adsorption at the solid-liquid interface and keep adsorbing, their ionic head groups facing the solution, lowering the interfacial tension. It is hypothesized that a hemimicelle structure forms at the PMMA-solution interface because the adsorption area of the surface-active agent on PMMA quickly reduces as the concentration of surface-active agent molecules increases. The surface tension remains unchanged at this point because of the saturation adsorption of surfactants at the air-liquid interface. However, when the interfacial tension drops, the contact angle quickly diminishes.

The third stage for C16 and C18 is the change from CMC to 1E-3. After the CMC, the surfactant's adsorption at the gas-liquid interface tends to reach saturation and the

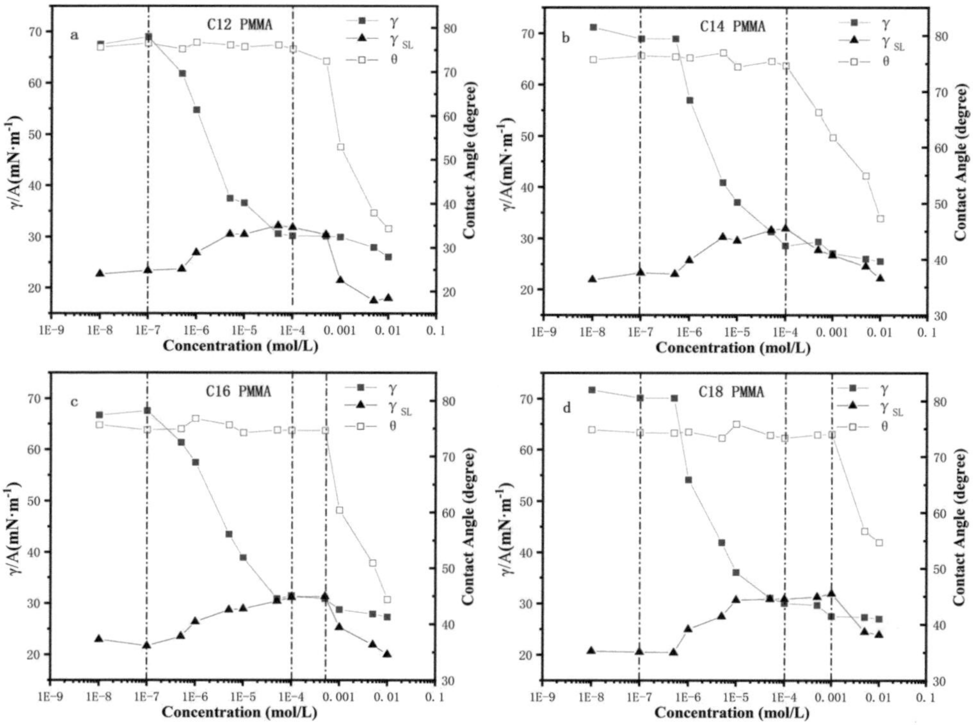

Figure 7.
(a–d) Concentration dependence of PMMA surface adhesion data for C12, C14, C16, and C18, respectively.

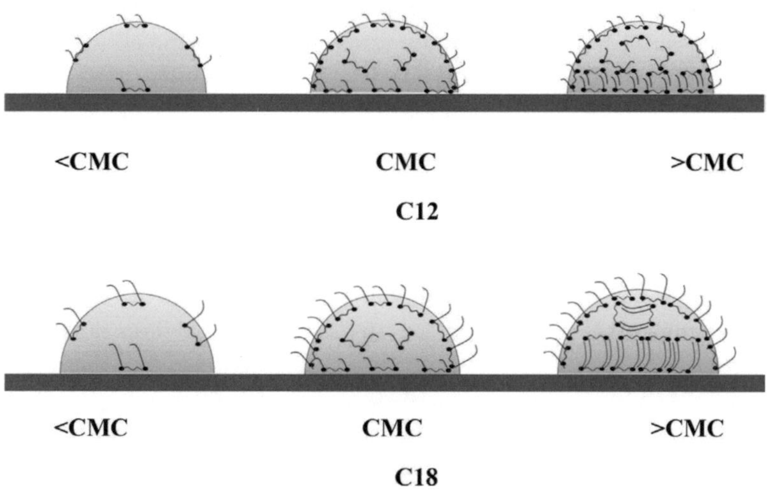

Figure 8.
Schematic diagram of the possible molecular arrangement of surfactant molecules on the surface of PMMA.

surface tension approaches a plateau. Since interfacial tension is still balanced, there is little change in the contact angle in the interim. The last step is 1E-2 to 1E-3. The surface tension tends to plateau in this phase following the CMC. But interfacial tension abruptly drops, causing the contact angle to drop precipitously. The adsorption behavior starts to alter because of surfactant molecules' growing preference for hydrophobic interactions to facilitate adsorption on the PMMA surface. As previously discussed, it is hypothesized that the adsorbed surfactants form aggregates at the solid-liquid interface due to the notable increase in surfactant molecule adsorption during this stage.

4. Conclusion

This paper focuses on the distinct wettability modification and adsorption behavior of Gemini anionic surfactants on the PMMA surface that have varying hydrophobic carbon chain lengths. These surfactants are synthesized from dialkyl dodecane succinic anhydride to form C12, dialkyl tetradecane succinic anhydride to form C14, dialkyl hexadecane succinic anhydride to form C16, and dialkyl octadecene succinic anhydride to form C18. The research explores how these Gemini surfactants interact with the PMMA surface, potentially enhancing its wettability and modifying its surface properties. The determination of surface tension shows that as the hydrophobic carbon chain length increases, the CMC value of the surfactant gradually decreases, and the γcmc first decreases and then increases. The four surfactants' contact angles hardly vary at low concentrations. After CMC treatment, the contact angle is significantly reduced. At high concentrations, C12 has the lowest contact angle. Calculations indicate that the four surfactants synthesized in this paper have a higher amount of surfactant molecules adsorbed at the PMMA-solution interface after passing through the CMC. Surfactants tend to form hemimicelles on the PMMA surface. The results indicate that the adsorption behavior of the four anionic Gemini surfactants synthesized in this paper on the PMMA surface is completely different from that of the monomers, and they exhibit lower surface tension values and CMCs. However, in terms of wetting behavior, the Gemini surfactant with the shortest hydrophobic chain, C12, shows a stronger wetting modification effect on the PMMA surface compared to the C14, C16, and C18 Gemini surfactants.

Acknowledgements

We gratefully acknowledge financial support from the National Key R&D Program of China (NO. 2019YFA0708700), TIPC Director's Fund of Technical Institute of Physics and Chemistry, Chinese Academy of Sciences. We would like to thank Xu Zhicheng and Jin Zhiqiang for their valuable suggestions during the experimental design phase. Also, we extend our thanks to Zhang Lei and Wang Chenguang for providing experimental materials and equipment support.

Author details

Zhe Hang[1,2], Lu Zhang[2], Fenrong Liu[1*] and Wangjing Ma[2*]

1 School of Material Science and Engineering, Inner Mongolia University, Hohhot, P. R. China

2 Technical Institute of Physics and Chemistry, Chinese Academy of Sciences, Beijing, P. R. China

*Address all correspondence to: fenrongl@163.com and wjma@mail.ipc.ac.cn

References

[1] Wang J, Gu F, Han W, et al. Green surfactant made from cashew phenol for enhanced oil recovery. ACS Omega. 2023;**8**(2):2057-2064

[2] Long X, He N, He Y, et al. Biosurfactant surfacing with pH-regulated emulsification activity for efficient oil separation when used as emulsifier. Bioresource Technology. 2017;**241**:200-206

[3] Ai C, Wang S, Liu C, et al. Experimental study on the influence of surfactants on ore surface wettability. ACS Omega. 2023;**9**(1):1056-1068

[4] Ramos GAR, Akanji LT, Afzal W. A novel surfactant–polymer/alkaline–surfactant–polymer formulation for enhanced oil recovery (eor) processes. Energy and Fuels. 2020;**34**(2):1230-1239

[5] Lai CC, Chen KM. Preparation and surface activity of polyoxyethylene-carboxylated modified gemini surfactants. Colloids and Surfaces A: Physicochemical and Engineering Aspects. 2008;**320**(1-3):6-10

[6] Lv WF, Zhou ZH, Zhang Q, et al. Wetting of polymer surfaces by aqueous solutions of branched cationic gemini surfactants. Soft Matter. 2019;**15**(33):6725-6731

[7] Shen YC, Jin Y, Lai S, et al. Synthesis, surface properties and cytotoxicity evaluation of nonionic urethane fluorinated surfactants with double short fluoroalkyl chains. Journal of Molecular Liquids. 2019;**296**:111851

[8] Jiang Q, Du Y, Zhang L, et al. Wettability of a polymethylmethacrylate surface by extended anionic surfactants: Effect of branched chains. Molecules. 2021;**26**(4):863

[9] Svanedal I, Andersson F, Hedenström E, et al. Molecular organization of an adsorbed layer: A zwitterionic, pH-sensitive surfactant at the air/water interface. Langmuir. 2016;**32**(42):10936-10945

[10] Yerliyurt K, Eğri S. Investigation on the potential use of polypropylene mesh for the reinforcement of heat-polymerized PMMA denture base resin. Polymers. 2022;**14**(16):3300

[11] Mani N, McLeod J, Sauder MB, et al. Novel use of polymethyl methacrylate (PMMA) microspheres in the treatment of infraorbital rhytids. Journal of Cosmetic Dermatology. 2013;**12**(4):275-280

[12] Saxena P, Shukla P. A comparative analysis of the basic properties and applications of poly (vinylidene fluoride) (PVDF) and poly (methyl methacrylate) (PMMA). Polymer Bulletin. 2022;**79**(8):5635-5665

[13] Miyamae T, Nozoye H. Morphology and chemical structure of poly (methyl methacrylate) surfaces and interfaces: Restructuring behavior induced by the deposition of SiO2. Surface Science. 2003;**532**:1045-1050

[14] Yuan T, Liu Z, Gao R, et al. Enhanced oil recovery from high-salinity reservoirs by cationic gemini surfactants. Journal of Applied Polymer Science. 2018;**135**(14):46086

[15] Hou B, Jia R, Fu M, et al. A novel high temperature tolerant and high salinity resistant gemini surfactant for enhanced oil recovery. Journal of Molecular Liquids. 2019;**296**:112114

[16] Hussain SMS, Kamal MS, Murtaza M. Synthesis of novel ethoxylated quaternary ammonium gemini surfactants for enhanced oil recovery application. Energies. 2019;**12**(9):1731

[17] Sanchez-Salazar E, Vazquez-Velez E, Uruchurtu J, et al. Use of a gemini-surfactant synthesized from the mango seed oil as a CO2-corrosion inhibitor for X-120 steel. Materials. 2021;**14**(15):4206

[18] Mahdavian M, Tehrani-Bagha AR, Alibakhshi E, et al. Corrosion of mild steel in hydrochloric acid solution in the presence of two cationic gemini surfactants with and without hydroxyl substituted spacers. Corrosion Science. 2018;**137**:62-75

[19] Mobin M, Noori S. Adsorption and corrosion inhibition behaviour of zwitterionic gemini surfactant for mild steel in 0.5 M HCl. Tenside Surfactants Detergents. 2016;**53**(4):357-367

[20] Zhang Q, Wang B, Xing X, et al. Wetting characteristics and kinetic behavior of amphoteric ionic surfactant on the microscopic "solid-liquid" interface of coal-a case study. Journal of Molecular Liquids. 2023;**392**:123497

[21] Wang C, Liu Y, Wang N, et al. Effects of catanionic surfactant mixture adsorption on the wettability of PTFE and PMMA. Colloids and Surfaces A: Physicochemical and Engineering Aspects. 2021;**631**:127659

[22] Chen XF, Xu ZC, Gong QT, et al. Adsorption of extended anionic surfactants at the water-polymethylmethacrylate interface: The effect of polyoxyethylene groups. Colloids and Surfaces A: Physicochemical and Engineering Aspects. 2023;**656**:130395

[23] Zhang D, Xu Z, Jin Z, et al. Wetting effect of branched anionic Gemini surfactant aqueous solution on PMMA surface. Soft Matter. 2023;**19**(24):4449-4457

[24] Du Y, Zhou ZH, Gao M, et al. Adsorption and wettability of extended anionic surfactants with different PO numbers on a polymethylmethacrylate surface. Soft Matter. 2021;**17**(26):6426-6434

[25] Ma WJ, Zhao S, et al. Preparation of Alkenyl Succinic Anhydrides. ZL201911131728.7. Beijing: China National Intellectual Property Administration; 23 Aug 2022

[26] Dix LR. Sodium salts of bis (1-dodecenylsuccinamic acids): A simple route to anionic gemini surfactants. Journal of Colloid and Interface Science. 2001;**238**(2):447-448

[27] Shi L, Ma J, Chen Y. Synthesis and surface properties of sodium sulfonate amphoteric surfactants having different hydrophobic carbon chain length. Research on Chemical Intermediates. 2022;**48**(6):2509-2533

[28] Szymczyk K, Jańczuk B. Wettability of polymeric solids by aqueous solutions of anionic and nonionic surfactant mixtures. Journal of Adhesion Science and Technology. 2011;**25**(19):2641-2657

[29] Ren HP, Tian SP, Zhu M, et al. Modification of montmorillonite by Gemini surfactants with different chain lengths and its adsorption behavior for methyl orange. Applied Clay Science. 2018;**151**:29-36

[30] Hu SS, Zhang L, Xu ZC, et al. Wettability alteration by novel betaines at polymer–aqueous solution

interfaces. Applied Surface Science. 2015;**355**:868-877

[31] Yuan FQ, Liu DD, Guo LL, et al. Effect of branched cationic and betaine surfactants on the wettability of a poly (tetrafluoroethylene) surface. Acta Physico-Chimica Sinica. 2015;**31**(4):715-721

Chapter 6

Study of Reducing and Adsorption Properties of Carbon Nanomaterials: Graphene and Hierarchically Structured Carbon Films – In Relation to Manganese Ions in an Aqueous Medium

Liudmila Yolshina and Varvara Dorogova

Abstract

A comparative study of the reducing and adsorption properties of TRGO, commercial graphene and HSCF synthesized in a salt melt on liquid magnesium during the interaction of carbon nanomaterials with sodium permanganate in a neutral medium was conducted. It was shown that all three carbon nanomaterials have a high reducing capacity—98.5% for TRGO, 99% for graphene and 100% reduction on the surface of HSCF. During the chemical interaction of TROG and HSCF, a layer of sodium manganite is formed on their surface. About 80% of the heptavalent manganese ions from a neutral sodium permanganate solution are reduced to the tetravalent state on the surface of both TROG and HSCF, and about 20% are reduced to the trivalent state. During the interaction of graphene, a nanocomposite material "graphene - manganese dioxide" is formed on its surface with manganese dioxide crystal sizes of up to 100 nm. Hierarchically structured carbon films synthesized from glucose on the surface of molten magnesium under a layer of molten salts have demonstrated high reducing and adsorption capacity during their interaction with sodium permanganate solution in neutral and alkaline media at temperatures from 20 to 80°C. It is shown that increasing the temperature from 20 to 80°C allows increasing the rate of permanganate ion reduction and adsorption of reduction products on carbon nanomaterials by 150 times; however, the interaction products do not depend on the process temperature. It was found that the synthesized HSCF with 100% reduction of permanganate ions in a neutral medium showed a high adsorption capacity, which is more than 450 mg of permanganate ions/g of HSCF and which is 1.5 times greater than the adsorption capacity of thermally reduced graphene oxide and 5 times greater than that of commercial graphene.

Keywords: graphene, thermally reduced graphene oxide, hierarchically structured carbon films (HSCF), permanganate ions, reduction, adsorption

1. Introduction

Groundwater pollution with heavy metal ions such as Mn (VII), Cr (VI), As (V), Pb (II), and others is a significant problem in regions with limited access to clean potable water. A lot of research has been done recently to find possible adsorbents that will remove heavy metal ions from the natural environment, including the removal of highly carcinogenic and mutagenic heptavalent manganese [1–3].

Sorption techniques are applied for final cleaning or when the concentration of pollutants is low. Both natural and synthetic materials with porous structures and active functional groups on the surface are used as sorbent materials [4, 5].

To purify wastewater from Mn (VII), various reducing agents and adsorbents can be used, which will allow Mn (VII) to be reduced to MnO_2 or Mn (II) depending on the acidity of the medium and adsorb the reduced form of manganese on its surface. The use of both functional hybrid nanomaterials—molybdenum sulfide MoS_2 [6], Co-Al-layered double hydroxide [7], mesoporous silicon dioxide [8], diatomite/MCM-41 composite [9], Foeniculum vulgare seeds [10], and natural adsorbents—is proposed as promising reducing agents and adsorbents of heptavalent manganese ions.

Various forms of carbon allotropes, primarily focusing on sp^2 hybridised carbon, such as graphene (G), graphene oxide (GO), and reduced graphene oxide (rGO) [11–13], are also promising adsorbents. It also highlights carbon nanotubes (CNT), which possess a high specific surface area and excellent adsorption properties. In addition to purifying polluted water, efforts are currently being made to produce hybrid composite materials based on graphene or its oxides combined with manganese dioxide or manganese spinel [14–16].

Different research groups have synthesised carbon composite materials consisting of carbon in various allotropic modifications. In addition to the synthesis of graphene, including that functionalised with multiple dopants, such as nitrogen, boron, sulfur, fluorine, or organic groups [17], sp^2/sp^3 or sp/sp^2 hybridised carbon hybrids are being developed: porous with a highly developed surface, which is considered as possible electrode materials for batteries or supercapacitors [18]. There is a modern approach to the synthesis of new carbon materials, such as porous films. The main idea is to form porous carbon films with a highly developed surface by thermal decomposition of organic products. In some cases, chemical treatment (e.g., with alkali solutions) was carried out [19].

Various methods have been proposed for synthesising novel carbon materials, such as porous films. The primary concept involves creating porous carbon films with a high surface area through the thermal degradation of certain materials, including soybeans [19], potatoes [20], eggplants [21], and pomelo peel [22]. In some instances, a chemical treatment with alkaline solutions is applied, followed by heating the semi-finished product at temperatures ranging from 200 to 1100°C [19]. Additionally, molten salt synthesis of carbon nanomaterial using zinc chloride as a reactive medium has been suggested [23]. In all these cases, a notable feature is that the resulting porous carbon structures achieve a very high specific surface area; however, they tend to be relatively thick and contain a significant amount of amorphous carbon with no regular structure.

Thus, the task of simultaneously purifying wastewater from toxic pollutants in the form of permanganate ions in a neutral environment using a cheap carbon nanomaterial with a high specific surface area and reducing and adsorption capacity and synthesising carbon-based composite materials with manganese dioxide crystals or its spinel formed on their surface, which can find application as a cathode material in chemical current sources [13, 14], a promising adsorption material for the extraction of ions of heavy polyvalent and rare earth metals from aqueous solutions [24–26], as well as organic pollutants [27, 28], seems extremely attractive.

The main methods for synthesising composite materials include reducing potassium or lithium permanganate in an acidic medium to create carbon-manganese dioxide or manganese spinel composites [29]. Other methods involve mixing manganese dioxide or permanganate with an organic precursor and annealing [15, 16]; electrodeposition on carbon nanotubes [30]; and radical polymerisation in an aqueous spinel suspension [31], and hydrothermal synthesis [32].

Our previous research focused on synthesising hierarchically structured carbon nanodiamond-graphene films by reacting glucose powder with molten salts and using a liquid metal (zinc or magnesium) at 700–750°C in air [33, 34]. This process produces large hierarchically structured carbon films (HSCF) with various carbon allotropes—sp, sp^2, and sp^3.

The reduction and adsorption properties of HSCFs were studied relative to hexavalent chromium ions in a neutral medium [35]. It was found that chromium reduction to the trivalent state and adsorption of Cr^{3+} ions from 0.05 M and 0.1 M solutions of $K_2Cr_2O_7$ occur within several seconds. The adsorption capacity of such carbon films reaches 111 mg/g at pH 7.

In this study, the synthesis and certification of nanocomposites "carbon nanomaterial (graphene, HSCF)- manganese dioxide or sodium-manganese spinel" was carried out with a direct chemical reduction of sodium permanganate on the surface of graphene or hierarchically structured carbon films with the deposition of nanocrystals of manganese dioxide or sodium-manganese spinel in dependence on the pH of the medium, the concentration of permanganate ions and the interaction temperature. The valence state of manganese in composite carbon-oxide materials was determined, and the effect of pH on the completeness of reduction and adsorption of the permanganate ion on the surface of the graphene and HSCF was studied.

HSCF synthesised by the interaction of glucose with molten magnesium under a layer of molten salts were used as the main object of the study. A comparison of the reducing and adsorption properties of the HSCF synthesised on magnesium was carried out with the properties of industrially produced graphene from BT Corp Generique Nano PVT Ltd. (India).

According to the data provided by the manufacturer [36], the graphene thickness is 1.6 nm, with 20% existing as a monolayer, 20% as bilayer graphene, and 60% as three-layer graphene. The positions of the Raman bands correspond to D 1351.7 cm^{-1}, G 1581.3 cm^{-1}, and 2D 2697.2 cm^{-1}. The peak ratios I_{2D}/I_G are 1.7054 and $I_D/I_G = 0.85$, indicating the formation of reasonably defective three-layer graphene. The specific surface area determined by the manufacturer using the BET method is small and amounts to 161 m^2/g. The graphene produced by BT Corp does not contain any carbonyl or carboxyl groups on its surface.

2. The synthesis and characterisation of hierarchically structured carbon films

The reaction of D-glucose with molten alkali metal halides was studied using liquid magnesium in a corundum crucible within a vertical furnace under air. Mixtures of lithium, sodium, potassium, and caesium chlorides, along with small amounts of ammonium, potassium, and aluminium fluorides (melting point below 700°C), were used. Details of the experimental technique are provided in [33, 34].

The melting point of magnesium is 658°C. Consequently, the selected reaction temperature was set to 100 degrees higher, so the reaction temperature was set to 750–800°C. At this range, small hydrogen bubbles formed with popping sounds resembling small flames. The process concluded when hydrogen release stopped, occurring after 0.5 to 2 hours, depending on the additives' composition.

X-ray analysis, XPS spectroscopy, Raman spectroscopy, and gas chromatography analysis of the reaction products show that the main products of such interaction are carbon, magnesium oxide and hydrogen. The proposed overall reaction equation, not accounting for the role of salt melts (alkali metal chlorides and fluorides), is as follows:

$$6Mg + C_6H_{12}O_6 \rightarrow 6C + 6MgO + 6H_2 \uparrow \quad (1)$$

$$\Delta G_{750^\circ} = -2558.78\, kJ/mol.$$

The change in the Gibbs energy of reactions with the use of magnesium is a large negative value. The process proceeds very vigorously with the release of hydrogen gas, the end of the release of which indicates the completion of the interaction.

The carbon films were examined immediately after the synthesis and adsorption of manganese ions using scanning electron microscopy and energy-dispersive spectroscopy by a TESCAN MIRA3 scanning electron microscope and an Inca Energy 250 EDS spectrometer. Raman spectra were recorded using a Renishaw U 1000 spectrometer coupled with a Leica DML microscope. An Ar^+ laser with a wavelength of 532 cm^{-1} and a power of 20mW was used as the excitation source. The specific surface of the HSCF before and after interaction with sodium permanganate solution was determined by the BET method using a SORBI N 4.1 device (Meta, Russia).

The surface of carbon films and carbon-oxide composites was studied using X-ray photoelectron spectroscopy by an X-ray photoelectron spectrometer KAlpha XPS System (Thermo Fisher Scientific). The obtained spectra were approximated using the Peak Fit software. The decomposition bands were analysed mainly using the Thermo Scientific XPS simplified database.

Micrographs of the obtained films with linear dimensions up to several mm are given in **Figure 1**.

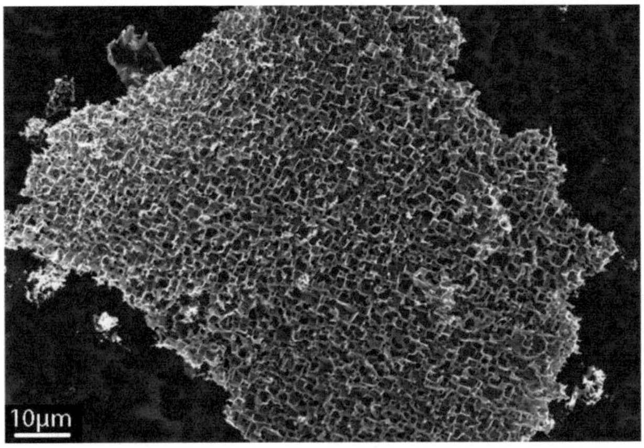

Figure 1.
Scanning electron microscopy images of front view of the film with composite hierarchically ordered crystal structure.

Figure 2.
Scanning electron microscopy images of: (a) the structured side and (b) the smooth side of the HSC films.

The films have two distinct sides, similar to those produced using zinc [33]. One side is smooth (**Figure 2a**), while the other has a complex hierarchical crystal structure (**Figure 2b**). The crystallites on the structured side are arranged in continuous steps, with a 90° angle between them. The average size of a crystallite measures approximately 2 μm, with a depth of about 1 μm. The EDS analysis of both the smooth and textured sides revealed that both consist of pure carbon. Higher magnification images show that the smooth surface is made up of carbon nanocrystallites, measuring up to 5 nm in size. The overall film thickness is around 1 μm.

The XRD patterns for the films collected in the 2θ range of 10 to 80° are similar to those for graphene (**Figure 3**) [37]. However, unlike typical graphene patterns, this one includes not only the diffuse lines from small carbon particles and amorphous carbon but also a prominent peak at 43.78 degrees, which corresponds to the (111) plane of cubic diamond.

The specific surface area of the carbon films is 290 m^2/g immediately after synthesis and rinsing with distilled water. This value rises to 547 m^2/g after washing with diluted hydrochloric acid, likely due to the removal of magnesium oxide nanoparticles formed during synthesis as indicated by reaction (1) [34]. For all subsequent studies, the carbon films were rinsed in hydrochloric acid, then distilled water, and dried in an oven.

X-ray photoelectron spectroscopy was used to determine the hybridised states of carbon atoms in the carbon films. The HSCF's survey spectrum (**Figure 4**) displays

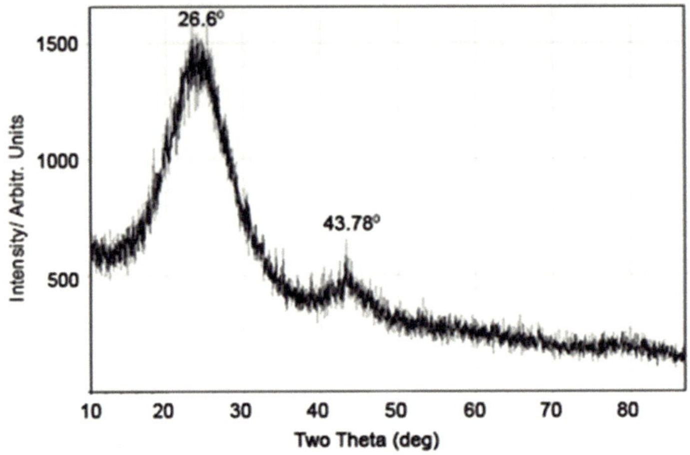

Figure 3.
X-ray powder diffraction pattern of HSC films.

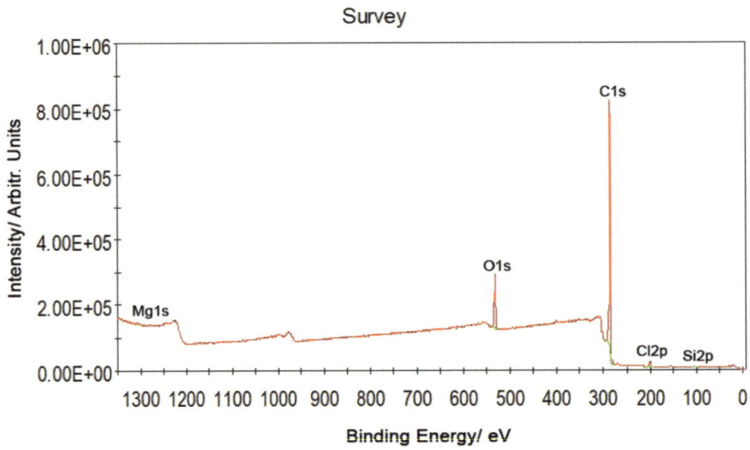

Figure 4.
XPS survey spectrum of HSC films.

Name	Peak BE	FWHM eV	Atomic %
Si2p	103.5	3.28	0.6
Cl2p	199.3	2.34	0.9
C1s	285.6	2.14	88.9
O1s	533.6	2.82	9.3
Mg1s	1304.6	2.27	0.3

Table 1.
Elemental ID and quantification of survey XPS spectrum of carbon film.

peaks of carbon and oxygen bound or adsorbed on its surface, comprising approximately 10%. It also shows trace amounts of magnesium and chlorine (**Table 1**).

In the high-resolution spectrum of carbon C1s (**Figure 5**), the main peak during deconvolution is a sharp and narrow peak at 284.4 eV, corresponding to carbon in sp^2 hybridisation, which indicates that about 40% of carbon atoms are in hexagonal graphene cells (**Table 2**). A carbon peak at 285.1 eV indicates diamond-like carbon in sp^3 hybridisation, while the peak at 284.0 eV shows that 13% of the carbon is in the sp-hybridised carbyne form. Thus, the HSCF produced contains a mixture of sp, sp^2, and sp^3 hybridised carbon atoms, but it is not merely another type of amorphous carbon, as it has a defined crystalline structure on one side.

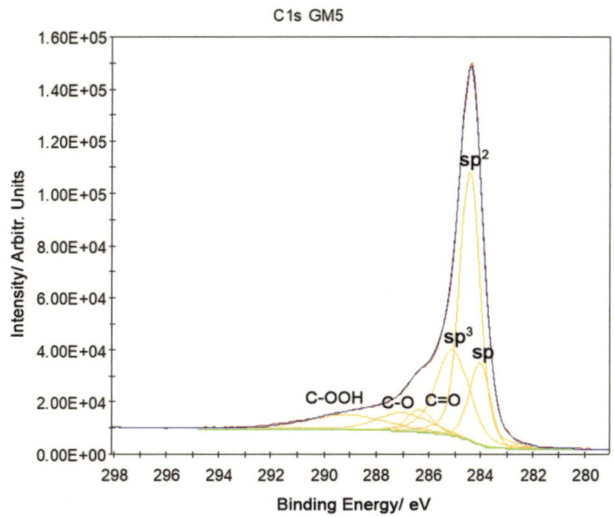

Figure 5.
XPS high-resolution spectrum of C1s HSC films.

Name	Peak BE	FWHM eV	Atomic %
C1s sp	284.0	0.95	11.7
C1s sp^2	284.4	0.91	38.9
C1s sp^3	285.1	1.53	20.7
C1s C=O	286.4	1.15	4.2
C1s C-O	287.0	2.38	6.9
C1s C-OOH	289.1	3.51	8.3
O1s	531.1	1.62	0.8
O1s	532.2	1.62	3.9
O1s	533.1	1.62	4.3
O1s	534.5	1.62	0.3

Table 2.
Elemental ID and quantification of high-resolution XPS spectrum of carbon film.

The high-resolution carbon spectrum shows peaks at 286.4, 287.0, and 289.1 eV, indicating C-O, C=O, and O-C=O bonds. These suggest the compound can reduce heavy metal ions in high oxidation states and may exhibit reducing properties in a neutral medium due to carbonyl and carboxyl group functionalisation. The proportion of carbon bound to oxygen is small and does not exceed 17%.

One of the most informative methods for analysing all carbon materials is Raman spectroscopy. The purity of carbon phases and their crystallinity are assessed by Raman spectroscopy, which is sensitive to a wide range of carbon structural states, including diamond, graphite, graphene, and fullerene. Certain Raman bonds are observed for each allotropic modification of carbon, allowing one to distinguish between sp, sp^2, and sp^3 hybridised states [38–40].

It is evident that the spectrum of the HSCF smooth side (**Figure 6a**) is similar to that of multilayer graphene according to the literature data; such spectrum is characterised by the broadening and shift of the D peak, decrease in the ID/IG ratio, and merging of the 2D and D + G peaks into one broad peak [38–40]. D peak is strongly shifted to the left as to the graphite peak (1355 cm^{-1}); there is a sharp peak on the top of this peak that corresponds precisely to cubic diamond (1332–1334 cm^{-1}), which has low broadening and sufficiently large intensity.

On the reverse side of the film, a sharp change is observed, namely, the complication of the spectrum (**Figure 6b**). In addition to the diamond peak (D) observed at 1332–1336 cm^{-1} and the graphene peak (G) at 1593–1600 cm^{-1} present across all

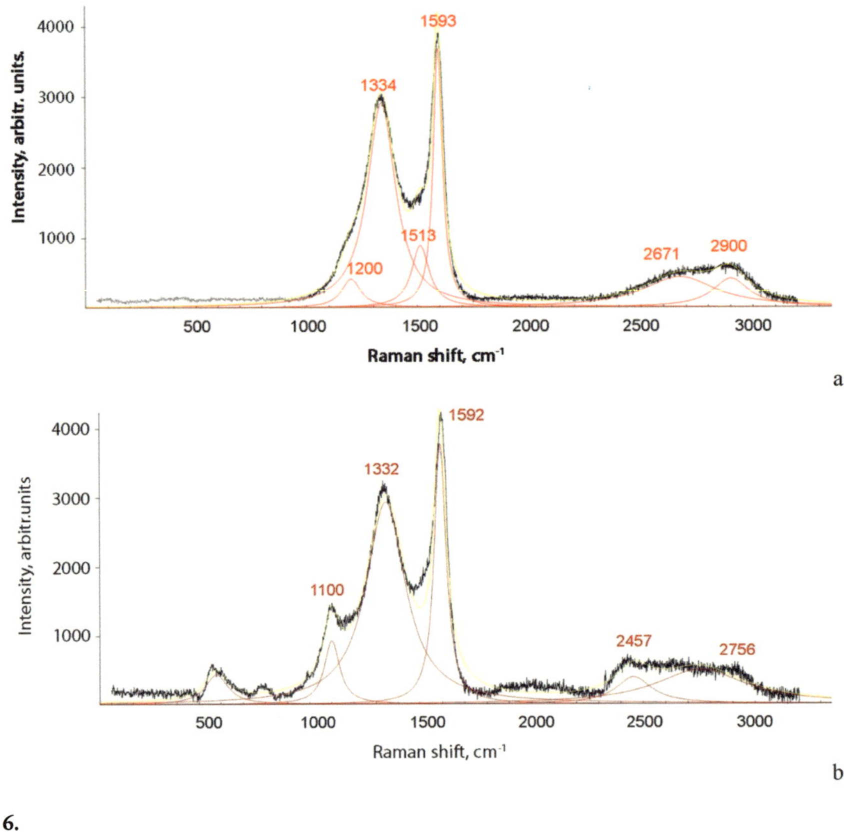

Figure 6.
Raman spectra of HSCF: a, smooth side; b, developed side.

areas of the film, there is also a peak in the visible region of the Raman spectra at 1120–1140cm^{-1} (**Figure 6b**). Several authors have initially attributed this peak to the presence of diamond clusters with an approximate diameter of 5 nm [38]. However, the band at 1140cm^{-1} and the broad band at 1480 cm^{-1} are close to the low-frequency and high-frequency bands of β-carbyne [41].

In the far region of the spectrum, a peak at 2450 cm^{-1} appears, which, according to the literature data [42], corresponds to the second-order cubic diamond.

Based on the data obtained by SEM, Raman and XPS spectroscopy, it can be concluded that the synthesised hierarchically structured carbon films are carbyne-graphene-nanodiamond films containing cumulene carbine-like chain carbon, multilayer graphene and sp^3-hybridised carbon with stress bonds (diamond).

3. Reduction and adsorption properties of carbon nanomaterials relative to the heptavalent manganese ions in aqueous medium at various pH values and temperatures

The work was carried out with acidified with the addition of hydrochloric acid to pH 1-6, neutral media at pH 7 and alkalised with the addition of sodium hydroxide to pH 8-14 with sodium permanganate solutions. The weighted portions of HSCF (0.2 g) were added to 20ml of sodium permanganate solution without any stirring at static mode at different concentrations and pH levels. The interaction time between the carbon film and sodium permanganate solutions at room temperature was 168 h (7 days), except for a 2-h exposure at different temperatures of 20, 40, 60, and 80°C for comparison.

The concentration of manganese and sodium in the solution before and after sorption was determined using an optical emission spectrometer OPTIMA 4300 DV (Perkin Elmer, USA) with inductively coupled plasma.

Carbon is known to exhibit the properties of a reducing agent. Its standard redox potential (E_0) is +0.475 V. Therefore, when MnO_4 - ions contained in the solution interact with a carbon nanocomposite, oxidation-reduction reactions can occur according to one of the following schemes:

$$3C^0 + 4MnO_4^- + H_2O = 3CO_3^{2-} + 2H^+ + 4MnO_2, \quad (2)$$

$$\Delta E^0 = 1.67 - 0.475 = 1.195\,V;$$

$$5C^0 + 4MnO_4^- + 2H^+ = 5CO_3^{2-} + H_2O + 4Mn^{2+}, \quad (3)$$

$$\Delta E^0 = 1.52 - 0.475 = 1.045\,V;$$

$$C^0 + 3H_2O + 4MnO_4^- = CO_3^{2-} + 6H^+ + 4MnO_4^{2-}, \quad (4)$$

$$\Delta E^0 = 0.54 - 0.475 = 0.065\,V.$$

According to the schemes of oxidation-reduction reactions (2–4), during the interaction of a hierarchically structured carbon film with a sodium permanganate solution at different pH, it is possible to reduce heptavalent manganese ions to Mn (II), Mn (IV), and Mn (VI) depending on the pH of the medium with the formation of poorly soluble precipitates—manganese dioxide or manganese spinel. Based on the values of standard oxidation-reduction potentials (ΔE_0) for the reactions of formation of MnO_2 and Mn^{2+} ions, there is a high probability of the formation of these substances in an aqueous solution during interaction with carbon since ΔE_0 for the formation of MnO_2 is 1.195 V, and for Mn^{2+} ions ΔE_0 = 1.045 V. The value of ΔE_0 for the reaction of the formation of MnO_4^{2-} ions under the same conditions is 0.065 V, which is significantly less than the two above-mentioned ΔE_0 values. Mn^{2+} ions will form in an acidic medium, while MnO_2 will form in neutral and alkaline conditions. Additionally, manganese dioxide crystals (manganese spinel) will chemisorb onto the carbon film.

The reduction and adsorption properties of carbon nanomaterials were studied in a static mode at temperatures from 20 to 80°C in sodium permanganate aqueous solutions with concentrations of 0.01, 0.02, 0.05, 0.06, 0.07, 0.08, and 0.1 M at pH 7 and 0.01 M at pH ranging from 1 to 14.

According to the chemical analysis, manganese and sodium concentrations in the filtered solution decreased significantly after the interaction with the HSCF sample due to reduction and sodium and manganese ions sorption at the surface of the carbon film.

The quantity of adsorbed ions was determined by assessing the difference in their concentration in the solutions before and after sorption. The adsorption value (mg/g) was calculated using the following formula (5):

$$a = \frac{c_{i-c_f}}{m} \times V, \tag{5}$$

where C_f and C_i denote final and initial concentrations of Mn ions, m is the sample of the sorbent, g; V is the volume of the analysed solution, ml.

The experiments aimed to determine how the purification of Mn ions in solutions varies with contact time with carbon material. **Figure** 7 provides the adsorption capacity and results of the Mn(VII) ions reduction at the surface of the HSC films, and it was found to be very efficient.

Further experiments were performed to determine the degree of reduction of Mn^{7+} ions in the solution, depending on the HSC sample weight. To determine the degree of reduction (R) of wastewater, use the formula based on the final (Cf) and initial (Ci) concentrations of Mn^{7+} ions:

$$R = \frac{c_i - c_f}{c_i} \times 100\%, \tag{6}$$

The amount of sodium permanganate adsorbed in equilibrium during 168 h at 25°C and pH 7 (ae, mg g^{-1}) at the surface of the HSC film as a function of the initial concentration of sodium permanganate is shown in **Figure** 7. An increase in sodium permanganate's initial concentration proportionally increases the carbon nanomaterial's adsorption capacity.

Figure 7 demonstrates that the reduction and adsorption of sodium permanganate occurs in direct proportion to its concentration in the aqueous solution, which is possible only in the case of the C-type isotherm according to the Giles classification. This case takes place when new adsorption centres appear. At the same time, the existing

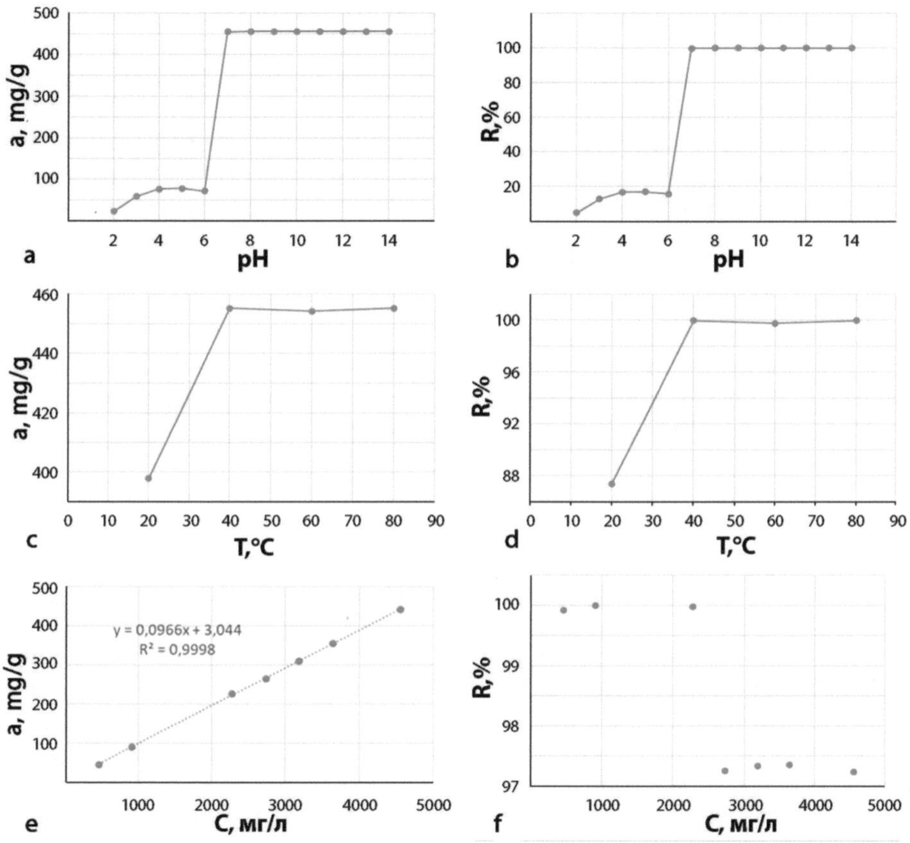

Figure 7.
Adsorption capacity (a, mg/g) and reduction degree of Mn^{7+} ions after the interaction of HSC films with the sodium permanganate solution as a function of pH (a, b), temperature (c, d), and sodium permanganate concentration (e, f).

adsorption centres are being filled, which increases the available adsorption surface in proportion to the amount of the substance adsorbed from the solution.

Due to the fact that we conducted studies using low concentrations of sodium permanganate, which prevented our curves from reaching saturation values. However, the data we obtained indicate that up to a $NaMnO_4$ concentration of 0.05 M, there is a complete (100%) reduction of heptavalent manganese ions. This reduction also leads to the deposition of MnO_2 or manganese spinel on the surface of the carbon adsorbent. A further increase in sodium permanganate concentration in the solution decreases the adsorption to 97.5%. The HSC film's adsorption capacity is high: about 450 mg/g of HSCF.

A study of the effect of pH on the HSC film—$NaMnO_4$ (0.01 M) interaction shows that in acidic solutions at pH values of 1-6, only about 10–20% of manganese ions are reduced and adsorbed, while at neutral and alkaline pH values of 7 to 14, a 100% reduction and adsorption of manganese ions at the surface of the HSC film is observed.

Thus, a 100% reduction of heptavalent manganese ions occurs on the surface of the HSCF, both in a neutral environment and in an alkaline environment, the completeness of which depends on the initial concentration of permanganate ions and the interaction temperature. At the same time, a high adsorption capacity of the carbon nanomaterial is achieved, which is more than 4500 mg/g when interacting with a 0.1 M sodium permanganate solution, which is an unprecedentedly high value for all known carbon materials.

An increase in the interaction temperature of HSC films and $NaMnO_4$ (0.01 M) from 20 to 40°C during 2 h increases the reduction and adsorption of manganese ions from 87 to 100%, respectively. A further increase in temperature neither increases nor decreases the adsorption properties. Consequently, the temperature increase positively affects the adsorption capacity of HSCFs concerning $NaMnO_4$, which indicates that adsorption, in this case, is the endothermic process.

Thus, a 100% reduction of heptavalent manganese ions occurs at the HSC surface both in neutral and alkaline media, the completeness of which depends on the initial concentration of permanganate ions and the interaction temperature. The high adsorption capacity of the carbon nanomaterial is achieved.

4. Structure and composition of HSC films after interaction with the $NaMnO_4$ aqueous solution at different pH values, temperatures and concentrations of sodium permanganate

The HSC films were analysed using physicochemical methods after interacting with the sodium permanganate solution.

After the interaction of the HSCF with the permanganate solution, regardless of the holding time, a significant change in the specific surface area occurs, namely, it decreases, with the most significant decrease in the specific surface area observed during the interaction of the ISUP at pH from 1 to 6, that is, in an acidic solution—by 20–30 times, and the more acid was added, the smaller the specific surface area of the film became (**Table 3**). **Figure** 7 shows that less than 20% of sodium permanganate is reduced and adsorbed onto the carbon surface in acidic solutions. Additionally, the interaction between the HSCF and sodium permanganate in both neutral and acidic solutions results in a reduction of the specific surface area of the carbon films, decreasing by a factor of 1.5 to 2. The most prominent specific surface area of the HSCF after the interaction was observed during the interaction at pH 8 (weakly alkaline solution) and with an increase in temperature to 40°C at pH 7 (**Table 4**). It is evident from the reduction-adsorption curves that in a neutral and alkaline environment, as well as in a neutral environment starting from 40°C, 100% reduction and

	Total specific surface (BET model, multipoint measurement), m^2/g
Initial HSC film	564.40 ± 7.76
pH2	20.7 ± 2.3
pH5	31.3 ± 3.5
pH7	315.7 ± 21.4
pH8	458.0 ± 3.5
pH14	296.4 ± 11.8

Table 3.
Total specific surface area of the initial HSCF and HSCF after interaction with $NaMnO_4$ *(0.01 M) at different pH values for 168 h.*

	Total specific surface (BET model, multipoint measurement), m^2/g
Initial HSC film	564.40 ± 7.76
T = 20 °C	271.7 ± 15.0
T = 40 °C	400.4 ± 3.8
T = 60°C	321.4 ± 16.7

Table 4.
Total specific surface area of the initial HSCF and HSCF after interaction $NaMnO_4$ (0.01 M) at pH 7 and different temperatures for 2 h.

adsorption of the interaction products occurs. Thus, it can be assumed that these conditions—interaction at neutral and slightly alkaline pH and an increase in temperature to 40–60°C—are optimal for the reduction-adsorption processes of heptavalent manganese on the carbon surface.

After 10 minutes of exposure to the 0.01 sodium permanganate solution, HSC films maintain their unique structure (**Figure 8**): developed and hierarchically structured on one side with the formation of numerous micro- and nanocrystals and smooth on the other side.

Increasing the sodium permanganate concentration in the solution and extending the exposure time of the HSC films result in the formation of multiple layers of crystals across the entire surface of the developed carbon film. In contrast, only individual crystals are formed on the smooth side of the film (**Figure 9**). That is, the HSC film is visually non-isomorphic, based on the ratio of different carbon hybridisations and the reactivity of the different sides.

According to the data provided in the table (**Figure 9**) and obtained using energy-dispersive X-ray spectra of the element concentrations at different points, the manganese and sodium concentrations are maximum in crystals.

Low-magnification microphotographs, as well as the X-ray element distribution maps, demonstrate that the HSC film—$NaMnO_4$ interaction product formed at pH7 is not uniform: there are places with higher concentrations of manganese, sodium,

Figure 8.
Microphotograph of HSC film after 10 min of interaction with sodium permanganate solution at pH 7.

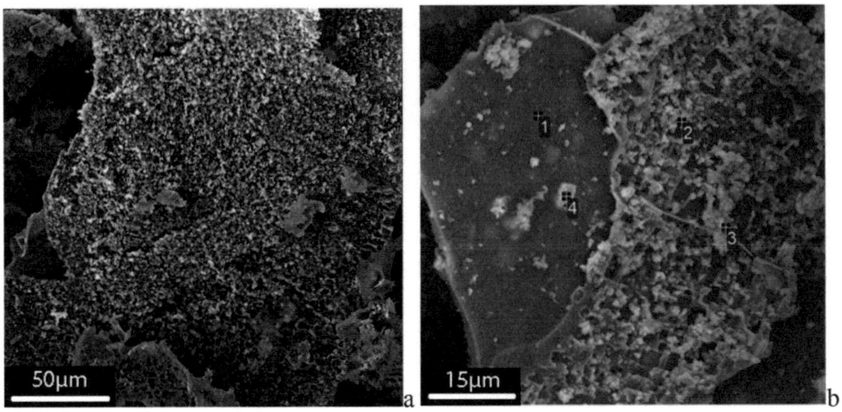

Elemental analysis of the crystals, wt. %

Spectrum	C	O	Na	Mn
1	58.26	27.50	0.59	13.78
2	35.18	37.45	0.68	26.78
3	15.73	31.96	1.02	51.25
4	30.60	33.51	0.76	34.58

Figure 9.
Microphotographs of the HSC film after interaction with the sodium permanganate solution at pH 7 for 2 h at 20°C: a, general view; b, electron photograph of the film with developed and smooth sides; c, table of element concentrations at different points of the film marked in ***Figure 9b****.*

and oxygen, and there are places where carbon is the main element of the films (**Figure 10**).

In addition, microphotographs and X-ray energy-dispersive spectra of the HSC films after interaction with the $NaMnO_4$ solution both in acidic and alkaline media and at temperatures increasing up to 80°C were obtained.

A lot of manganese is adsorbed on the surface of the carbon film (**Figure 11**) after the interaction of HSCF with permanganate ions in acidic media. Still, the oxygen content does not exceed a few percent, which is too low for manganese oxide formation, and no sodium content is observed. Obviously, when interacting in the acidic medium with the $NaMnO_4$ solution, the surface loses its hierarchical structure. Thus, as expected, the interaction in an acidic solution proceeds according to reaction (3) with a reduction to divalent manganese ions; the interaction proceeds without forming any additional phase on the surface of the carbon film.

Microphotographs of the HSC film after the interaction with the $NaMnO_4$ solution (0.01 M) at pH 14 (**Figure 12**) demonstrate that numerous microcrystals are formed only on the developed side during the interaction in an alkaline medium. Sodium content in such films is 5–10 times higher than in those obtained at neutral pH.

Thus, we can assume that the HSCF—$NaMnO_4$ solution (0.01 M) interaction in the acidic medium changes the film structure; it flattens, and manganese ions adsorb at the surface and do not form any additional phases. In neutral and alkaline media, the HSCF—$NaMnO_4$ solution interaction leads to the formation of numerous crystals on the developed film surface, while another side remains smooth with individual, separate crystals.

The surface of HSCF after interaction with 0.01 M $NaMnO_4$ solution at pH7 with an increase in the interaction temperature to 40°C (**Figures 13** and **14**) remains a carbon film with a high degree of filling of the developed surface with numerous crystals of sodium manganite. Since the entire surface is covered with crystals, they continue

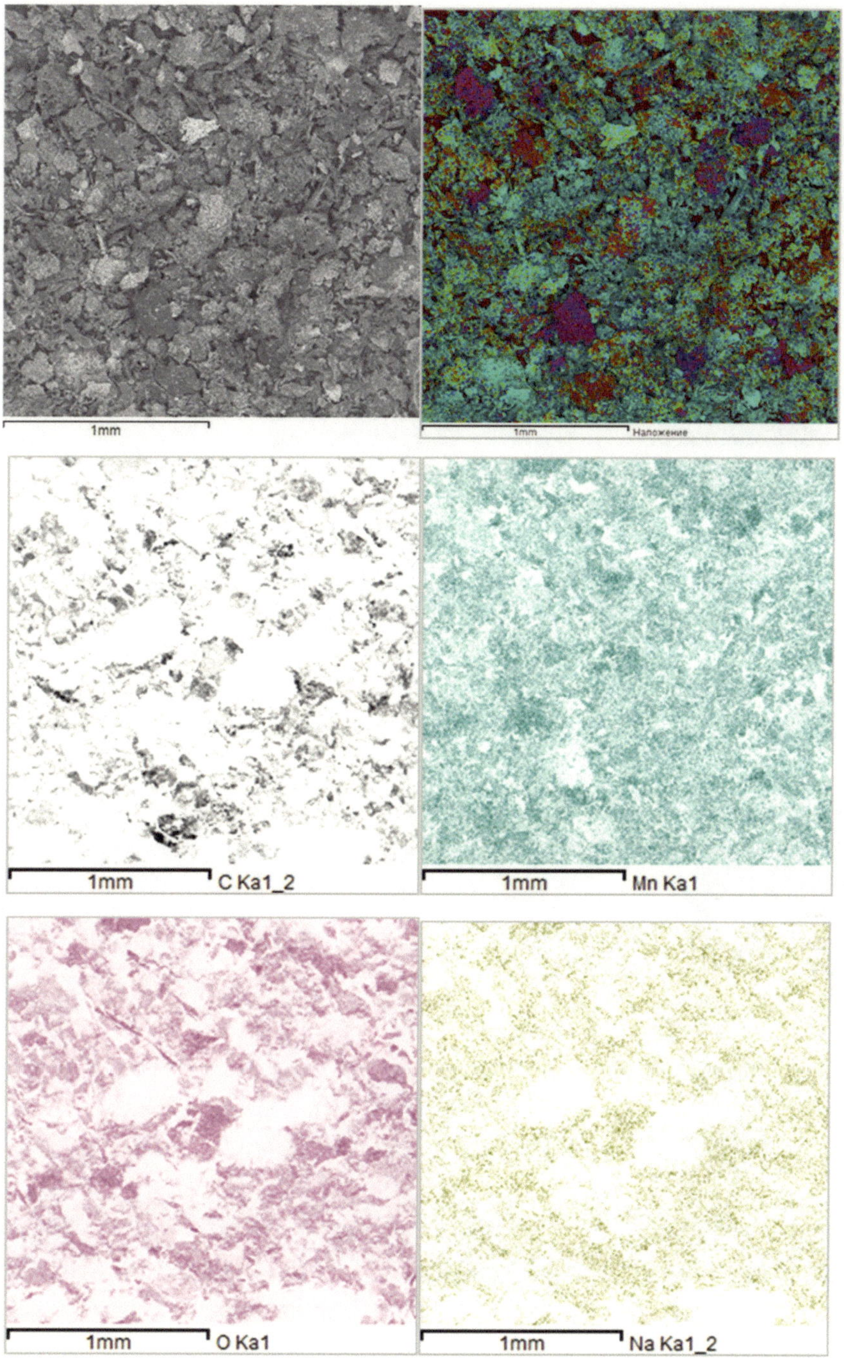

Figure 10.
Element distribution maps in an HSC film after interaction with sodium permanganate.

to precipitate into the already occupied positions of other crystals, forming complex structures that are in agreement with the adsorption isotherms. They show that an increase in temperature to 40°C leads to 100% adsorption of manganese from the solution. Further increase in temperature up to 80°C does not lead to any significant

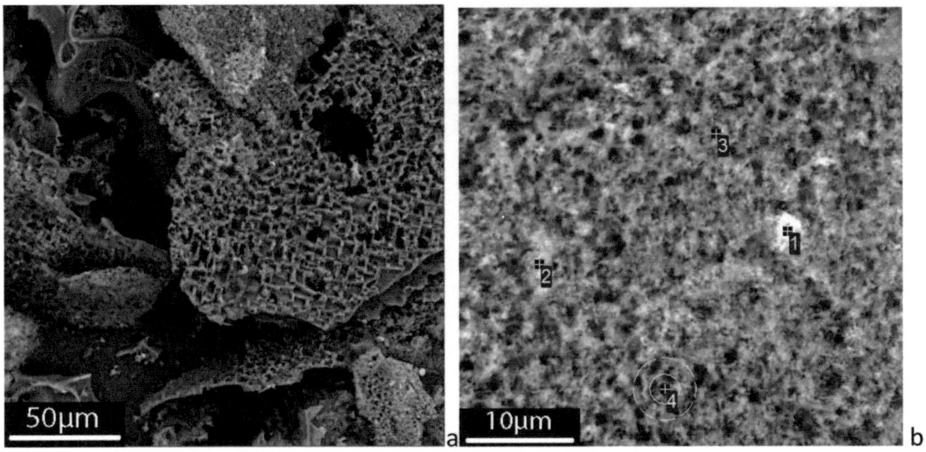

Spectrum	C	O	Mn
1	45.65	2.37	47.62
2	63.85	4.13	27.33
3	58.98	3.19	33.01
4	60.15	1.75	32.19

Figure 11.
Microphotographs of the HSC film after interaction with the sodium permanganate solution at pH 2 for 2 h at 20°C: a, general view; b, electron photograph of the film from the developed side; c, table of element concentrations at points marked in ***Figure 11b****.*

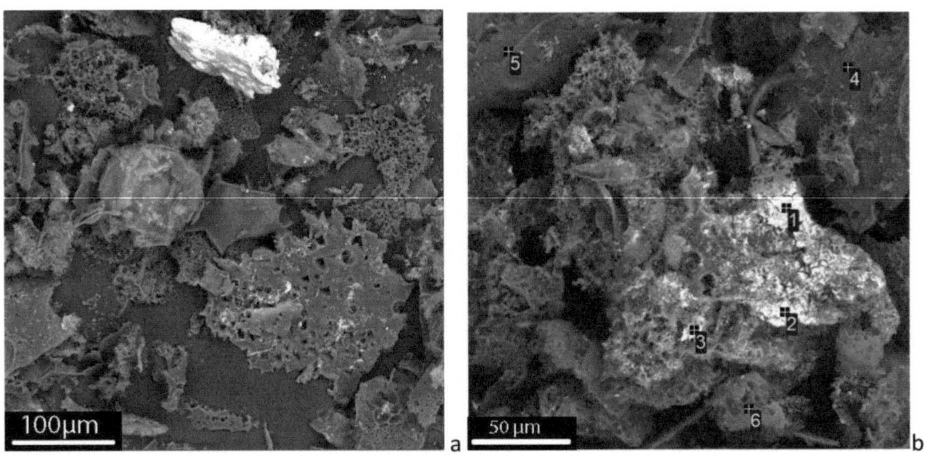

Spectrum	C	O	Na	Mn
1	23.0	13.8	3.5	56.6
2	35.0	24.3	5.2	32.8
3	58.2	20.3	3.1	16.8
4	88.1	9.5	0.6	1.5
5	92.7	5.9	0.4	

Figure 12.
Microphotographs of HSC film after interaction with the sodium permanganate solution for 2 h at pH 14 and 20°C: a, general view; b, electron photograph of the film from the developed side; c, table of element concentrations at points marked in ***Figure 10b****.*

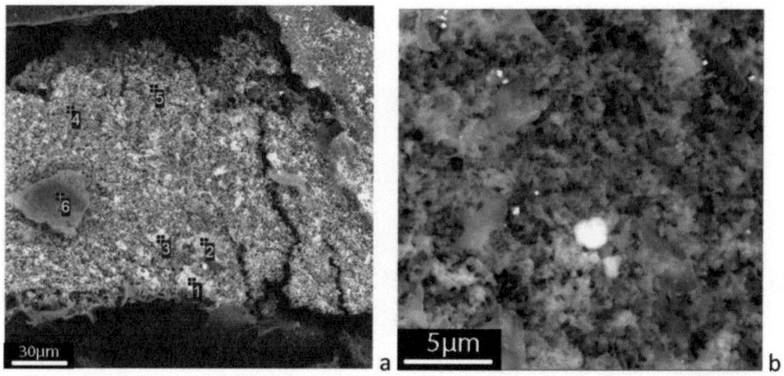

Spectrum	C	O	Mg	Si	Cl	Ca	Mn
1	62.46	20.03	1.11	0.25	0.17	0.18	14.96
2	62.53	16.17	1.02	0.22	0.27	0.23	19.57
3	57.60	21.78	1.30	0.23	0.21	0.24	18.66
4	60.21	19.03	1.12		0.26	0.27	19.12
5	59.32	20.18	1.42	0.24			18.84
6	83.01	10.49	2.24		2.78		1.47

Figure 13.
Microphotographs of the HSC film after interaction with the sodium permanganate solution for 2 h at pH 7 and 40°C: a, general view; b, electron photograph of the film from the developed side; c, table of element concentrations at points marked in ***Figure 13b****.*

Figure 14.
Microphotographs of HSCF after interaction with sodium permanganate solution for 2 h at pH 7 and 80°C: high power electron photograph of the film.

changes, as numerous crystals are deposited at the HSC film surface, and, as can be clearly seen in **Figure 14**, the oxide crystals deposit at the surfaces of the previously formed crystals. The adhesion of oxide crystals to the carbon film is very high, and the crystals remain at the surface of the carbon film after filtration and rinsing in distilled water.

5. XRD and Raman spectroscopy analysis of the product resulting from the HSC film and sodium permanganate interaction

The product of the interaction of the HSCF with a 0.01 M $NaMnO_4$ solution (pH 7) at room temperature and after annealing at 800°C was also studied by X-ray diffraction in the 2θ range of 0.0001 ~ 89.7223° with a signal accumulation time of 1 sec. Typical diffraction patterns are shown in **Figure 15**. The general appearance of the X-ray diffraction pattern obtained after the interaction at room temperature and pH 7 (**Figure 15a**) is similar to the X-ray diffraction pattern of the original HSCF synthesised on magnesium (**Figure 3**). However, in addition to the diffuse carbon peaks, a small peak at 18°, a fairly large peak at 36°, a small peak at 38°, and minor peaks at 44 and 65° are observed. These values are very close to the standard peaks of sodium manganite.

To confirm the formation of spinel $NaMn_2O_4$, the product, after the interaction of HSCF with 0.01 M $NaMnO_4$ solution at pH 7, was dried and annealed for 4 h at 800°C to remove all carbon. The obtained X-ray diffraction pattern (**Figure 15b**) fully corresponds to the formation of spinel $NaMn_2O_4$ with the space group Fd3m

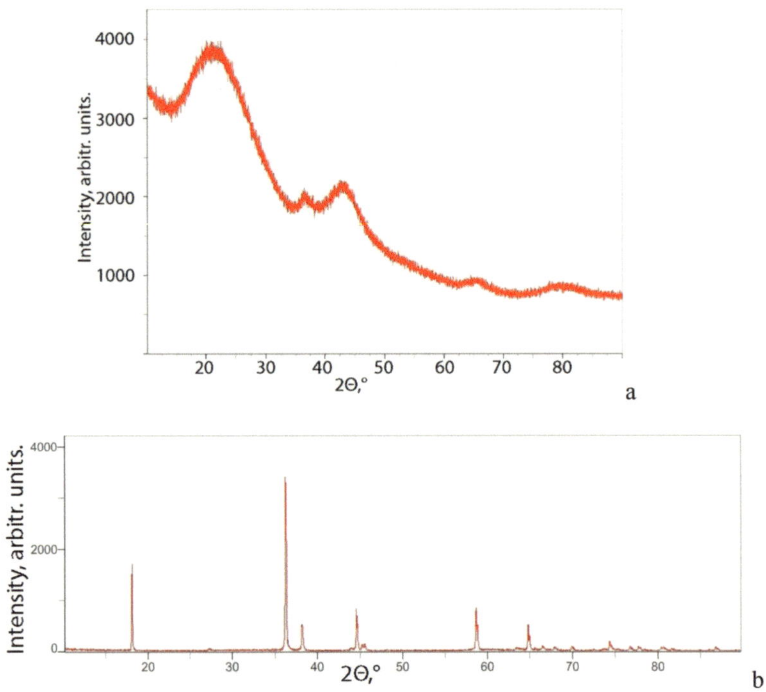

Figure 15.
X-ray diffraction pattern of the product of interaction of HSCF with 0.01 M $NaMnO_4$ при pH 7 solution at: (a) room temperature; after annealing in the air at 800°C during 4 h (b).

with characteristic peaks at 18.03° (111), 36.12° (311), 38.10° (222), 44.46° (400), 58.57° (511) and 64.69° (440). Thus, X-ray phase analysis proved that the product of the interaction of HSCF with the sodium permanganate solution is spinel $NaMn_2O_4$. However, due to the difficulties in determining the interaction product by X-ray phase analysis due to the need for additional annealing stages, it was decided to further define the interaction product by the express and accurate method of Raman spectroscopy.

Raman spectroscopy determined the interaction product of the HSC film—$NaMnO_4$ (0.01–0.1) solution interaction at temperatures ranging from 20 to 80°C and pH values ranging from 1 to 14.

Raman spectra (**Figure 16**) recorded at different points of the developed and smooth sides of the hierarchically structured film after interaction with the 0.01 M $NaMnO_4$ solution at 20°C at pH 7 show that in addition to D at 1330–1340 cm^{-1} and G at 1580–1590 cm^{-1} peaks characteristic to all carbon nanomaterials, there is a peak at ~630–640 cm^{-1}. This peak is almost undefined on the smooth side, while on the developed side, the 630 cm^{-1} peak is dominant. The height of this peak relative to D and G carbon peaks increases as the spectrum is recorded at the most crystal-filled places of the carbon film, where the ratio of the 630 cm^{-1} to 1340 cm^{-1} peak intensities is more than 10 (blue curve, **Figure 16**). According to the literature data, it is known that the dominant peak at 630 cm^{-1} occurs due to symmetrical twisting vibrations of the Mn-O bond in MnO_6 octahedrons (A1g mode) [43, 44] and characterises the formation of spinel $NaMn_2O_4$.

The Raman spectra of the HSC film (**Figure 17**) after the interaction with the 0.01 M $NaMnO_4$ solution in the acidic pH region did not determine the formation of sodium manganite, which is confirmed by microphotographs (**Figure 11**); no crystals are formed at the carbon surface, while manganese is effectively adsorbed in the form of ions (table to **Figure 11**).

Raman spectra from the alkaline pH region indicated the formation of sodium manganite on the carbon surface. The peak intensity decreased with increasing pH,

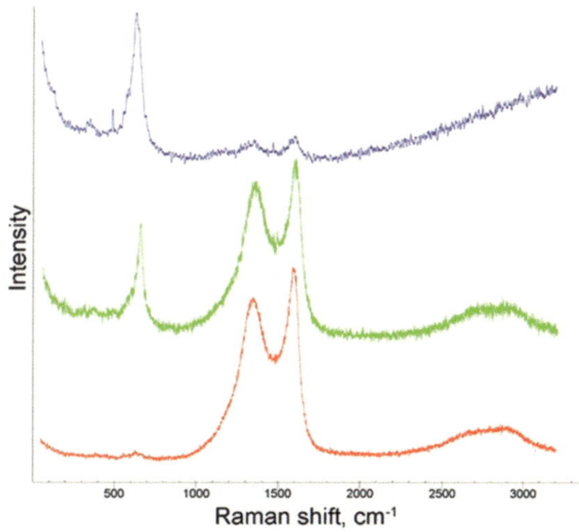

Figure 16.
Raman spectra of the HSC film after interaction with the $NaMnO_4$ solution (0.01%) at 20°C and pH 7: red denotes the smooth side, green denotes the developed side of the film with crystals, blue denotes the spectrum of the crystal clustering site.

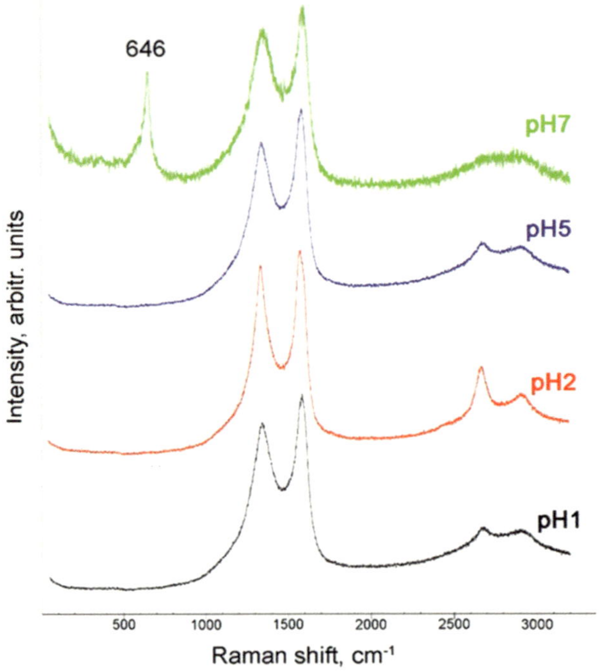

Figure 17.
Raman spectra of the HSC film after interaction with the $NaMnO_4$ (0.01%) solution at 20°C in the acidic pH region.

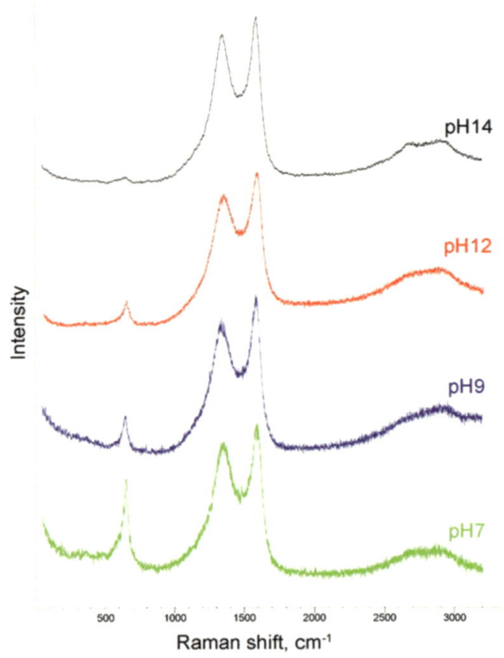

Figure 18.
Raman spectra of HSCF after the interaction with the $NaMnO_4$ (0.01%) solution at 20°C in the alkaline pH region.

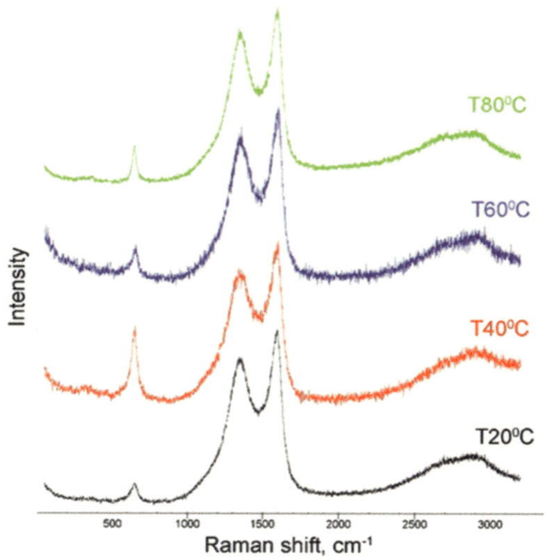

Figure 19.
Raman spectra of HSCF after the interaction with the $NaMnO_4$ (0.01%) solution at different temperatures and the pH 7 value.

likely due to poor adhesion of the products to the carbon substrate and the shedding of oxide phases during analysis (see **Figure 18**).

As seen from Raman spectra (**Figure 19**), the temperature growth to 40–80°C increases the ratio of the MnO_6 group peak to carbon peaks.

It should be noted that in the Raman spectra of the HSCF after interaction with a sodium permanganate solution at all pH values and temperatures studied, the peak with a wavelength of 1140 cm^{-1}, which, in our opinion, refers to sp-hybridised carbon in the form of β-carbyne, disappears, as well as the peak at 2450 cm^{-1}, which is the peak of second-order diamond. When interacting with sodium permanganate in various environments, the structure, morphology, and valence state of the HSCF undergoes significant changes.

Thus, it can be considered that the optimal conditions for the synthesis of the HSCF manganese spinel nanocomposite are synthesis from an aqueous solution with direct interaction of the HSCF with a sodium permanganate solution in a neutral or slightly alkaline medium—up to pH 9 at a temperature of 40–60°C.

6. X-ray photoelectron spectroscopy analysis of the product resulting from the HSC film and sodium permanganate solution interaction

The survey photoelectron spectrum of the HSC film after exposure to the $NaMnO_4$ solution illustrates the following peaks: dominant C 1s, O 1s, Na1s and Mn 2p (**Figure 20**). Residual Cl and Mg traces, remaining in the carbon film after rinsing from chloride melt after synthesis, are detected. **Tables 5–7** provide the main components of the film determined after HSCF—0.01 M $NaMnO_4$ solution interaction in neutral, acidic and alkaline media with the peak's resolution.

The carbon concentration in the film after HSCF—$NaMnO_4$ solution interaction at pH 7–14, as calculated from the photoelectron spectrum, is approximately 50 atomic percent (at. %). The oxygen concentration reaches about 33 at. %, while the

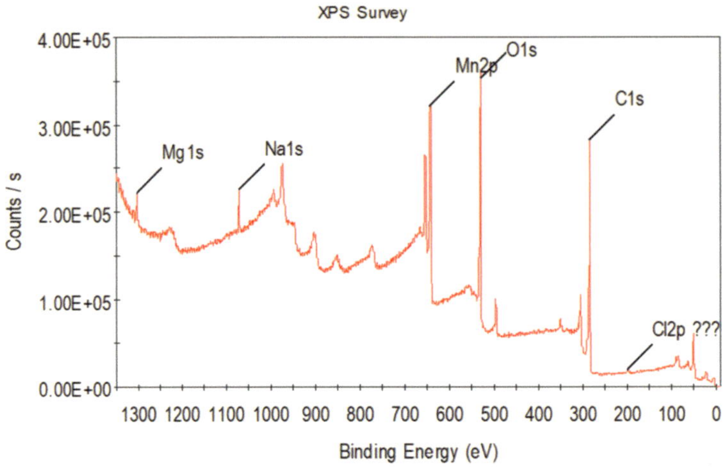

Figure 20.
Survey photoelectron spectrum of HSC film after the interaction with the sodium permanganate solution at pH 14.

Name	Peak BE	Height CPS	FWHM eV	Atomic %
C1s sp^2	284.15	21484.51	0.89	45.14
C1s sp^3	284.75	6174.61	0.9	1292
C1s C = O	285.7	3647.08	1.99	16.83
C1s C-OOH	288.07	1551.88	3.01	10.74
O1s	531.07	1211.26	1.41	1.63
O1s	532.48	5914.7	2.25	12.75

Table 5.
Chemical analysis of the HSC films after the interaction at the pH 2 value according to the XPS.

Name	Peak BE	Height CPS	FWHM eV	Atomic %
Mn3s	84.19	1807.69	3	0
Mn3s	88.87	1248.43	3	0
C1s sp^2	284.38	30440.73	1.13	32.21
C1s sp^3	285.71	5871.2	1.81	10.02
C1s C-OOH	288.07	2487.05	3	7.04
O1s MeO	529.69	48874.95	1.09	20.66
O1s A	530.89	10423.84	1.67	6.78
O1s B	532.09	8130.29	1.67	5.29
O1s C	533.29	4645.92	1.67	3.03
Mn2p3/2 (III)	640.7	11010.36	1.54	2.55
Mn2p3/2 A	641.95	27622.73	1.28	5.32
Mn2p3/2 B	642.83	13362.18	1.28	2.57
Mn2p3/2 C	643.58	11748.88	1.28	2.26

Name	Peak BE	Height CPS	FWHM eV	Atomic %
Mn2p3/2 D	644.77	5701.56	1.28	1.1
Mn2p sat E	646.66	3193.27	3	0.96
Na1s	1071.14	589.4	1.8	0.21

Table 6.
Chemical analysis of the HSC films after the interaction at the pH 7 value according to XPS.

Name	Peak BE	Height CPS	FWHM eV	Atomic %
Mn3s	83.97	721.09	3	
Mn3s	88.38	787.88	2.53	
C1s sp^2	284.16	17117.09	0.92	29.71
C1s sp^3	284.76	4870.13	0.92	8.46
C1s C = O	285.58	2895.32	1.79	9.78
C1s C-OOH	288.06	1303.43	3.01	7.35
O1s MeO	529.48	17401.91	1.03	14
O1s B	530.91	7719.36	2.08	12.53
O1s C	532.9	2927.02	2.08	4.76
Mn2p3/2 (III)	640.89	6526.59	1.49	2.93
Mn2p3/2 (IV)A	641.9	8801.08	0.96	2.55
Mn2p3/2 (IV)B	642.7	6098.83	1.05	1.94
Mn2p3/2 (IV)C	643.35	3082.94	0.86	0.8
Mn2p3/2 (IV)D	644.05	2203.65	1.06	0.7
Mn2p3/2 (IV)E	644.85	850.92	0.98	0.25
Mn2p3/2 (IV)F	645.9	685.95	1.5	0.31
Na1s	1071.03	3735.05	1.55	2.23

Table 7.
Chemical analysis of the HSC film after the interaction at the pH 17 value according to XPS.

manganese concentration ranges from 13 to 20 at. %. Additionally, the introduction of extra sodium in the form of NaOH to alkalinise the $NaMnO_4$ solution—raising the pH from 8 to 14—results in a natural increase in sodium content in the manganese spinel crystals, increasing from 0.4 at. % at pH 7 to 5 at. % at pH 14.

However, according to the chemical analysis data, the HSC film adsorbs about 20 wt.% of manganese in the range of acid solutions. According to the X-ray energy-dispersive analysis, a manganese concentration of about 5 at.% is detected at the surface of the HSC film after the interaction with the $NaMnO_4$ solution acidified with hydrochloric acid solution. The general XPS spectrum of the HSC film after the interaction in the acidic environment at pH 1-6 does not contain a peak of manganese.

The high-resolution C1s spectra of the HSC film before and after the interaction with the sodium permanganate solution at pH ranging from 1 to 14 (**Figure 21**) were compared. The main difference in the HSCF's carbon spectra before and after the interaction with Mn^{7+} ions is the fact that the impact of sp-hybridised carbon with a binding energy of 238.8 eV completely disappears from the spectra after the interaction. At the same time, its content in the initial HSC film was about 12%. The diamond sp^3-hybridised carbon content also decreases by about 10%, while the

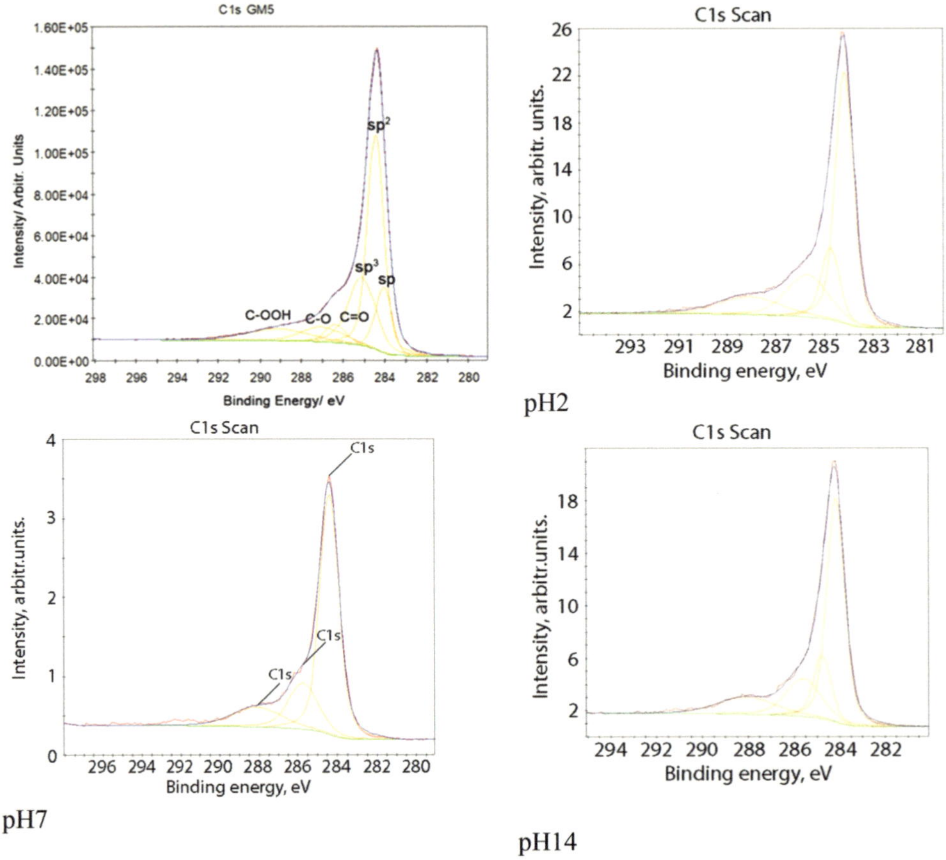

Figure 21.
High-resolution XPS spectra of C1s: a, initial HSC film and HSC film after the interaction with the $NaMnO_4$ solution at pH: b, 2; c, 7; d, 14.

carbon content included in the carbonyl C=O and carboxyl groups C-OOH increases. The content of sp^2-hybridised carbon in the film rises after the interaction in an acid medium and slightly decreases after the interaction in neutral and alkaline media.

Thus, an examination of the high-resolution XPS spectra of C1s carbon indicates that sp-hybridised carbon plays a crucial role in the reduction of manganese ions, leading to the formation of manganese spinel crystals on the surface of HSCF. This carbon form is entirely consumed during the chemical reaction, demonstrating that sp-hybridised carbon is the most reactive among the three hybridisation states of carbon.

Determination of the Mn valence state by the photoelectron spectra is usually performed within the Mn2p region. As can be seen from **Figure 22b**, the most suitable approximation was chosen with the parameters [45] for Mn^{4+} with the impact of Mn^{3+}.

Another way of identifying the Mn oxidation degree is to evaluate the Mn3s region, which has two multiple cleavage components. The value of the cleavage peak is a diagnostic sign of the degree of magnesium oxidation.

$$\Delta E \text{ for } MnO\left(Mn^{2+}\right) = 6.0\,eV,\ Mn_2O_3\left(Mn^{3+}\right) \geq 5.3\,eV,\ MnO_2\left(Mn^{4+}\right) = 4.7\,eV$$

The difference between the peaks is also closer to MnO_2 and equals 4.6 eV (**Figure 22c**).

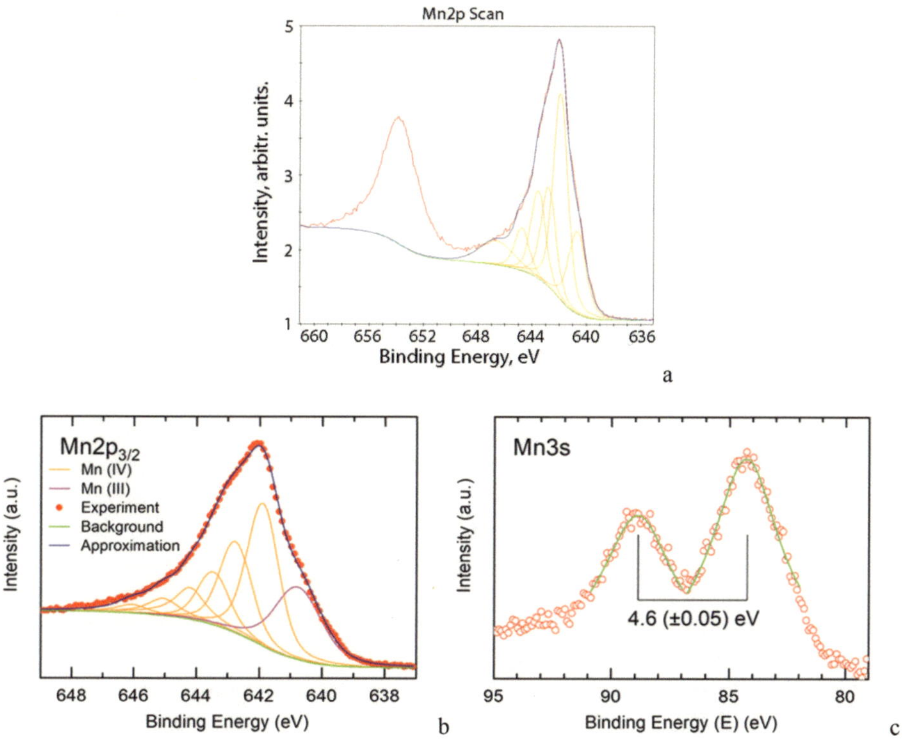

Figure 22.
Typical high-resolution XPS spectrum of Mn2p (a, b) with approximation and HSC film spectrum Mn3s (c) after interaction with the sodium permanganate solution at pH 7.

Calculation of the reduced manganese content in the trivalent and tetravalent states at the HSC film surface demonstrates that approximately 20% of manganese is present in the trivalent state during its interaction with sodium permanganate solution in both neutral and alkaline media.

Manganese spinel nanocrystals mainly deposit on the developed side of the carbon film due to the absence of the sp-hybridised carbon peak in the HSCF spectra. The maintained step-like structure on this side suggests that sp^2 and sp^3 hybridised carbon primarily contributes to it.

The different reactivity of carbon depending on its valence states has been proved for the first time.

A complete 100% Mn reduction and adsorption from sodium permanganate solutions (0.01 M–0.1 M) was observed in both neutral and alkaline environments. The hierarchically structured carbon film demonstrated an impressive adsorption capacity of over 450 mg/g when interacting with the 0.1 M sodium permanganate solution. This capacity is one of the highest reported among all previously studied carbon nanomaterials.

7. Study of the products of reduction and adsorption of permanganate ions by the graphene

This section compares the reducing and adsorption activity of industrial graphene and HSCF with sodium permanganate in a neutral environment.

Commercial graphene powder (**Figure 23**) is actually a multilayer graphene, as expected from the estimate of its specific surface area, which is 20 times smaller than the specific surface area of single-layer graphene. In addition, the Raman spectrum of industrial graphene also had all the characteristic bands—D at 1351 cm^{-1}, G at 1581 cm^{-1} and 2D at 2697 cm^{-1}, but the intensity ratio of the I_{2D}/I_G bands is less than 0.3, indicating the formation of graphene with more than five layers. Graphene powder was used in further synthesis experiments without additional preparation procedures.

Weighed samples of the HSCF and graphene (0.02, 0.05, 0.08, and 0.1 g) were added to a sodium permanganate solution (20 mL), and the resulting solutions stayed in static mode. The process occurs very slowly at room temperature and finishes within 336 h. The temperature of 60°C (at which the synthesis prolongs within 2 h) was chosen for further studies.

The data on the contents of manganese ions in the solution before and after 2 h of the interaction of the HSCF and graphene with a neutral solution of sodium permanganate are given in **Table 8**. It is clear that manganese adsorption on the HSCF is more effective. With an HSCF additive of 0.1 g, manganese is almost completely extracted from the solution. In contrast, when using the same amount of graphene, only a tenfold decrease in manganese concentration is observed. Increasing the amount of graphene further does not enhance its ability to adsorb manganese on its surface.

The specific surface area and energetic properties of carbon nanomaterials significantly influence the efficiency and kinetics of reduction and adsorption processes in a neutral medium. This relationship underscores the importance of these surface characteristics in determining the overall performance of carbon nanomaterials in various applications.

The specific surface area of the HSCF is three times greater than that of commercial graphene, which has a value of 161 m^2/g, while the HSCF measures 547 m^2/g. This significant difference results in faster and more complete adsorption on the HSCF.

After interaction with 0.01 M sodium permanganate solution at temperatures of 20 and 60°C, graphene was analysed using scanning electron microscopy,

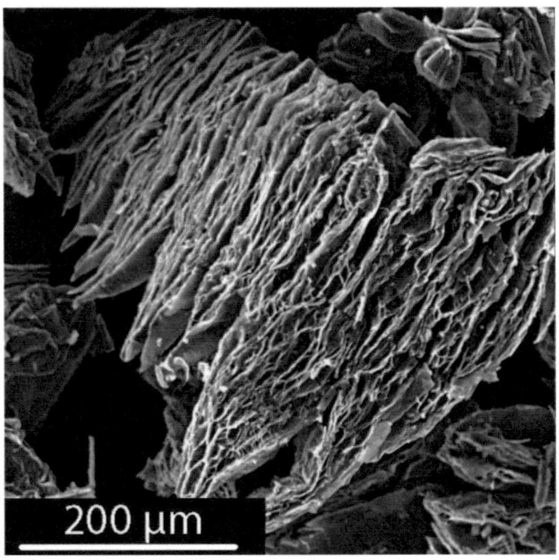

Figure 23.
Microphotograph of a commercial graphene powder produced by the BT Corp Generique Nano PVT.Ltd.

HSCF, weighted sample, g	Mn ions, mg	Graphene, weighted sample, g	Mn ions, mg
0	550	0	550
0.02	0.3	0.02	28
0.05	0.05	0.05	2.8
0.08	0.025	0.08	3.4
0.1	0.013	0.1	3.6

Table 8.
Contents of potassium and manganese ions in the solution before and after interacting with the carbon nanomaterials.

energy-dispersive analysis, X-ray phase analysis, X-ray photoelectron and Raman spectroscopy. It was shown that the composition and morphology, as well as the dimensionality of the newly formed phases, do not depend on the interaction temperature, so in what follows, all analysis data will be presented for samples obtained at a temperature of 60°C.

The microphotographs of graphene after interacting with a 0.01 M solution of $NaMnO_4$ (pH 7) are shown in **Figure 24**. It is obvious that nanocrystals cover all surfaces, not just the active edges of graphene, including interlayer ones. The XRD pattern of such a composite (**Figure 25**) contains peaks of γ-MnO_2 (the main phase), as well as graphite (substrate) and impurity β-MnO_2. Numerous crystals of manganese dioxide are well-faceted and have a size of up to 100 nm.

The Raman spectrum of this composite (**Figure 26**) contains all characteristic graphene peaks: D at 1350 cm^{-1}, G at 1583 cm^{-1}, and 2D at 2703 cm^{-1}. In addition, sharp, pronounced peaks are observed at 566.7 and 641 cm^{-1}, corresponding to different modifications of manganese dioxide [28].

The survey XPS spectrum of commercial graphene after interaction with sodium permanganate at pH 7 (**Figure 27**), in addition to the peaks of carbon at 284.83 eV, oxygen at 530 eV, sodium at 1071.3 eV, manganese at 642.5 eV also contains peaks of elements that appeared in graphene during the production process, namely, chlorine, nitrogen, calcium and zinc (**Table 9**). It should be noted that the sodium content in the oxide phase is less than the sodium content in the HSCF in a 0.01 M permanganate solution at neutral pH. The Raman spectrum and X-ray diffraction pattern indicate the synthesis of manganese dioxide instead of sodium manganite during the interaction of graphene with a sodium permanganate solution.

In the high-resolution spectrum of Mn2p (**Figure 28b**), in addition to tetravalent manganese, up to 18 at. % of trivalent manganese is determined (**Table 10**), which is somewhat lower than for HSCF. The peak difference of Mn3s gives a value of 4.48 eV, which correlates well with the data for MnO_2 (Mn^{4+}), where the peak difference is 4.7 eV. The carbon peak C1s in the high-resolution spectrum (**Figure 28a**) mainly consist of the sp^2 hybridised state; graphene does not change its electronic state when interacting with the sodium permanganate solution.

Taking into account the data obtained by various methods, it can be concluded that when interacting with a solution of sodium permanganate in a neutral environment at temperatures of 20 and 60°C, a nanocomposite material "graphene-manganese dioxide" is formed on graphene. As a result of such interaction, crystals of predominantly γ-MnO_2 with a size of no more than 100 nm are deposited on the entire surface of graphene. The specific surface of the "graphene -manganese dioxide" composite decreases more than threefold and is 45.2 m^2/g.

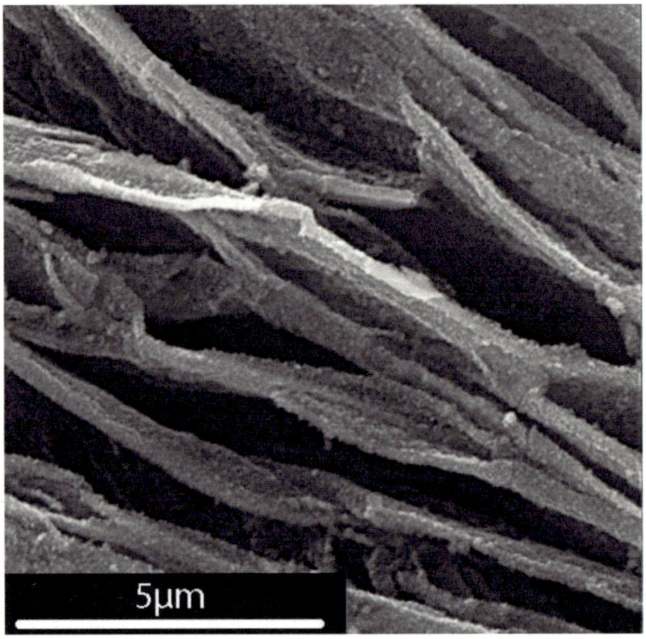

Figure 24.
Microphotograph of graphene after interacting with a 0.01 M solution of $NaMnO_4$.

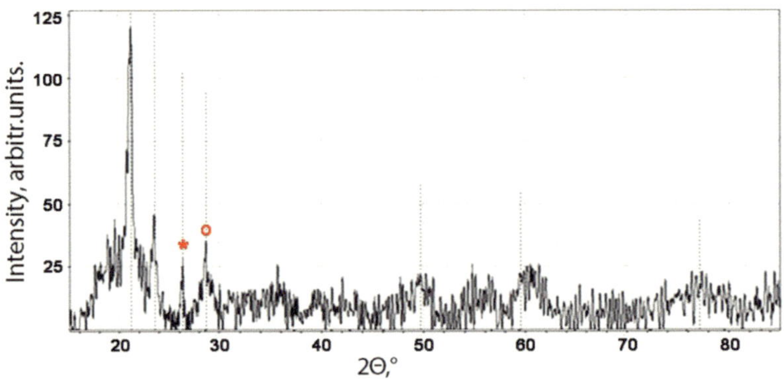

Figure 25.
*XRD pattern of the graphene–MnO_2 nanocomposite: *graphite; °β, MnO_2. Rest peaks correspond to γ-MnO_2.*

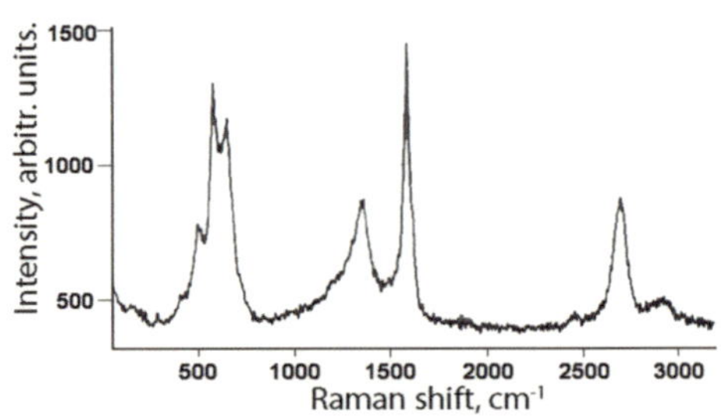

Figure 26.
Raman spectrum of the graphene–MnO_2 nanocomposite.

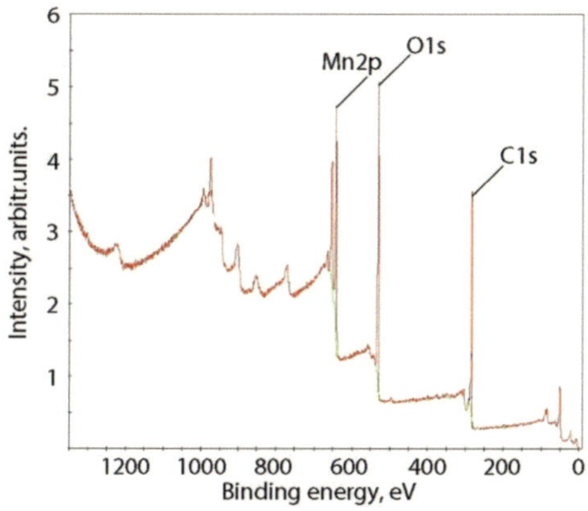

Figure 27.
Survey XPS spectrum of commercial graphene after interaction with sodium permanganate at pH 7.

Element	Peak position, eV	Peak height, rel.units	FWHM eV	Atomic %
Cl2p	198.72	4297	4.23	0.61
C1s	284.72	304,376	1.99	48.95
K2p	292.59	5602.57	1.27	0.16
Ca2p	347.33	11757.33	2.41	0.83
N1s	399.82	7210.29	3.66	1.69
O1s	529.99	408600.1	2.19	33.49
Mn2p	642.52	309,327	3.12	12.97
Na1s	1071.34	7356.25	1.92	0.32
Zn2p	1044.75	14317.06	1.43	0.97

Table 9.
Identification of elements and quantitative estimation of the survey XPS spectrum of commercial graphene after interaction with Mn(VII) ions.

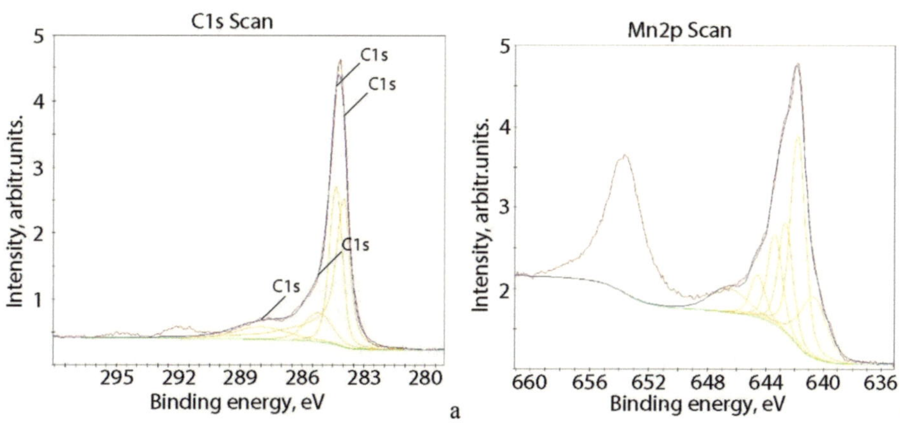

Figure 28.
XPS high-resolution spectra of commercial graphene after interaction with sodium permanganate at pH: a, C1s; b, Mn2p.

Element	Peak position, eV	Peak height, rel.units	FWHM eV	Atomic %
Mn3s	84.06	1777.75	3	0
Mn3s	88.54	1194.25	3	0
C1s	284	22545.84	0.83	24.31
C1s	284.37	23987.36	0.74	16.09
C1s	285.22	4446.27	1.57	6.31
C1s	287.97	1878.36	3	5.12
O1s MeO	529.42	51113.41	1.12	21.34
O1s A	530.62	8055.17	1.79	5.39
O1s B	531.82	6464.02	1.79	4.33
O1s C	533.12	5319.68	1.79	3.57
Mn2p3/2 (III)	640.7	7552.8	2.18	2.38
Mn2p3/2 A	641.8	25898.52	1.23	4.62
Mn2p3/2 B	642.66	13472.82	1.23	2.4
Mn2p3/2 C	643.41	11002.6	1.23	1.96
Mn2p3/2 D	644.6	5217.87	1.23	0.93
Mn2p sat E	646.58	2983.35	3	0.86
Na1s	1070.85	959.25	2.12	0.38

Table 10.
Identification of elements and quantitative estimation of the high-resolution XPS spectrum of graphene after interaction with Mn (VII) ions.

8. Conclusions

A method for synthesising a new nanocarbon material by reacting D-glucose with liquid active metals (Mg, Zn) in a molten mixture of alkali chlorides at a temperature of 750°C has been developed, resulting in the formation of uniformly dispersed in the salt melt free-standing films consisting of 100% carbon, up to 1 μm thick with a great specific surface area of up to 590 m^2/g. It was established by the means of physico-chemical analysis that the formed films have a complex hierarchical structure and consist of carbon in sp, sp^2, and sp^3 hybridised states, which correspond to β-carbyne, graphene, and cubic (or hexagonal) diamond, respectively. One side of the film is smooth, consisting predominantly of sp^2 hybridised carbon. The other has a developed surface on which diamond crystallites are arranged in the form of continuous steps with an angle between them equal to 90°.

Carbon nanomaterials, including hierarchically structured carbon films (HSCF) and commercially produced graphene, have been found to be effective in decontaminating water from heptavalent manganese ions [46–48]. When these carbon nanomaterials react with a permanganate solution in a neutral medium, the resulting products can vary. Specifically, graphene typically produces manganese dioxide, whereas hierarchically structured carbon films may yield manganese spinel or manganese oxide.

A comparative study of the reducing and adsorption properties of commercial graphene and HSCF with sodium permanganate in a neutral medium was conducted.

It was shown that both carbon nanomaterials have a high reducing capacity—99% for graphene and 100% reduction on the surface of HSCF.

During the chemical interaction of HSCF, numerous layers of sodium manganite nanocrystals were predominately formed on the developed side of their surfaces. 80% of the heptavalent manganese ions from a neutral sodium permanganate solution are reduced to the tetravalent state on the surface of HSCF, and about 20% are reduced to the trivalent state. During the interaction of graphene with permanganate ions in neutral media, a nanocomposite material, "graphene - manganese dioxide," is formed with manganese dioxide crystals of up to 100 nm.

Hierarchically structured carbon films (HSCF) demonstrated complete 100% manganese (VII) ion reduction and adsorption activity during their interaction with sodium permanganate solution in neutral and alkaline media at temperatures ranging from 20 to 80°C. The temperature growth from 20 to 80°C allows a 150-fold increase in the rate of reduction and adsorption of heptavalent manganese on HSCF; however, the interaction products do not depend on the process temperature.

The hierarchically structured carbon film's adsorption capacity is more than 450 mg/g of Mn^{7+} ions during the interaction with the sodium permanganate (0.1 M) solution, which is one of the highest achieved adsorption capacities of all previously studied carbon nanomaterials in neutral media.

The chemical interaction of HSCF with $NaMnO_4$ in neutral and alkaline media results in the formation of sodium-manganese spinel nanocrystals (up to 100 nm long and up to 10 nm thick) on the developed side of the hierarchically structured carbon film. At the same time, manganite crystals are practically not formed on the smooth side of the film, which indicates the different reactivity of the sides of the HSC films consisting of pure carbon and having different structures and morphology.

Acknowledgements

The work was performed at the Institute of High Temperature Electrochemistry of the Ural Branch of the Russian Academy of Sciences within the framework of the grant of the President of the Russian Federation SP-6070.2021.1. Facilities of the Shared Access Centers "Composition of Compounds» and Ural Center for Shared Use "Modern nanotechnology" Ural Federal University were used.

Author details

Liudmila Yolshina* and Varvara Dorogova
Institute of High-Temperature Electrochemistry UB RAS, Ekaterinburg, Russia

*Address all correspondence to: yolshina06@rambler.ru

References

[1] Wuan RA, Okieimen FE, Imborvungu JA. Removal of heavy metals from a contaminated soil using organic chelating acids. International journal of Environmental Science and Technology. 2010;7:485-496

[2] Li L, Li Y, Yang C. Chemical filtration of Cr (VI) with electrospun chitosan nanofiber membranes. Carbohydrate Polymers. 2016;**140**:299-307. DOI: 10.1016/j.carbpol.2015.12.067

[3] Fu F, Wang Q. Removal of heavy metal ions from wastewaters: A review. Journal of Environmental Management. 2011;**92**:407-418

[4] Sun J, Zhang Z, Ji J, Dou M, Wang F. Removal of Cr^{6+} from wastewater via adsorption with high-specific-surface-area nitrogen-doped hierarchical porous carbon derived from silkworm cocoon. Applied Surface Science. 2017;**405**:372-379

[5] Srivastava SK, Tyagi R, Pant N. Adsorption of heavy metal ions on carbonaceous material developed from waste slurry generated in local fertilizer plants. Water Research. 1989;**23**(9):1161-1165

[6] Ghobadi S, Samiey B, Ghanbari A. Adsorption and reduction coupling of permanganate on MoS_2: Water treatment and metal ion separation. Journal of Solid State Chemistry. 2021;**304**:122588

[7] Ghobadi S, Samiey B, Esmaili E, Cheng C-H. Comparison of kinetics of adsorption of permanganate on Co-Al-layered double hydroxide and MoS_2 nanocompounds. Acta Chimica Slovenica. 2023;**70**:44-58

[8] Chen F, Hong M, You W, Li C, Yu Y. Simultaneous efficient adsorption of Pb^{2+} and cmO_4^- ions by MCM-41 functionalized with amine and nitrilotriacetic acid anhydride. Applied Surface Science. 2015;**357**:856-865

[9] Ahmed IA, Badawi M, Bonilla-Petriciolet A, Lima EC, Seliem MK, Mobarak M. Insights into the Mn(VII) and Cr(VI) adsorption mechanisms on purified diatomite/MCM-41 composite: Experimental study and statistical physics analysis. Frontiers in Chemistry. 2022;**9**:814431. DOI: 10.3389/fchem.2021.814431

[10] Bani-Atta SA. Potassium permanganate dye removal from synthetic wastewater using a novel, low-cost adsorbent, modifed from the powder of Foeniculum vulgare seeds. Scientifc Reports. 2022;**12**:4547

[11] Kr R, Verma R, Kapoor S, Gupta K, Chaudhari RR. An efficient technique for removal of K^+ and MnO_4^- ions through adsorption in aqueous solution by using activated charcoal. The Pharmaceutical and Chemical Journal. 2014;**1**(2):20-25

[12] Egbosiuba TC, Abdulkareem AS, Kovo AS, Afolabi EA, Tijani JO, Roos WD. Enhanced adsorption of As(V) and Mn(VII) from industrial wastewater using multi-walled carbon nanotubes and carboxylated multi-walled carbon nanotubes. Chemosphere. 2020;**254**:126780

[13] Kumar N, Rodriguez JR, Pol VG, Sen A. Facile synthesis of 2D graphene oxide sheet enveloping ultrafine 1D $LiMn_2O_4$ as interconnected framework to enhance cathodic property for Li-ion battery. Applied Surface Science. 2019;**463**:132-140

[14] Chen L, Li D, Zheng X, Chen L, Zhang Y, Liang Z. Integrated

nanocomposite of $LiMn_2O_4$/graphene/carbon nanotubes with pseudocapacitive properties as superior cathode for aqueous hybrid capacitors. Journal of Electroanalytical Chemistry. 2019;**842**:74-81

[15] Kumar R, Soam A. Synthesis and characterization of sucrose derived carbon/MnO_2 nanocomposite. Materials Today: Proceedings. 2021;**35**:76-78

[16] Tabassam R, Alvi F, Aslam N, Raza R, Saifur-Rehman LS, et al. Electrochemical investigation of $LiMn_2O_4$/asphalt and $LiMn_2O_4$/bituminous coal based cathode composites for efficient lithium-ion battery. Materials Letters. 2021;**302**:130275

[17] Fierro S, Watanabe T, Akai K, Einaga Y. Highly sensitive detection of Cd2+ using boron doped diamond electrodes. Electrochimica Acta. 2011;**82**:9-11. DOI: 10.1016/j.electacta.2012.03.030

[18] Bettini LG, Della Foglia F, Piseri P, Milani P. Interfacial properties of a carbyne-rich nanostructured carbon thin film in ionic liquid. Nanotechnology. 2016;**27**(11):115403. DOI: 10.1088/0957-4484/27/11/115403

[19] Long C, Jiang L, Wu X, Jiang Y, Yang D, Wang C, et al. Facile synthesis of functionalized porous carbon with three-dimensional interconnected pore structure for high volumetric performance supercapacitors. Carbon. 2015;**93**:412-420. DOI: 10.1016/j.carbon.2015.05.040

[20] Fuertes AB, Sevilla M. High-surface area carbons from renewable sources with a bimodal micro-mesoporosity for high performance ionic liquid-based supercapacitors. Carbon. 2015;**94**:41-52. DOI: 10.1016/j.carbon.2015.05.040

[21] Li Z, Lv W, Zhang C, Li B, Kang F, Yang Q-H. A sheet-like porous carbon for high-rate supercapacitors produced by the carbonization of an eggplant. Carbon. 2015;**92**:11-14. DOI: 10.1016/j.carbon.2015.02.054

[22] Liu J, Li H, Zhang H, Liu Q, Li R, Li B, et al. Three-dimensional hierarchical and interconnected honeycomb-like porous carbon derived from pomelo peel for high performance supercapacitors. Journal of Solid-State Chemistry. 2018;**257**:64-71. DOI: 10.1016/j.jssc.2017.07.033

[23] Deng X, Zhao B, Zhu L, Shao Z. Molten salt synthesis of nitrogen-doped carbon with hierarchical pore structures for use as high-performance electrodes in supercapacitors. Carbon. 2015;**93**:48-58

[24] Kano N, Pang M, Deng Y, Imaizumi H. Adsorption of rare earth elements (REEs) onto activated carbon modified with potassium permanganate ($KMnO_4$). Journal of Applied Solution Chemistry and Modeling. 2017;**6**:51-61

[25] Gao Z, Xu S, Li L, Yan G, Yang W, Wu C, et al. On the adsorption of elemental mercury on single-atom TM (TM = V, Cr, Mn, Co) decorated graphene substrates. Applied Surface Science. 2020;**516**:146037

[26] Mopoung S, Bunterm T. $KMnO_4$ modified carbon prepared from waste of pineapple leaf fiber production processing for removal of ferric ion from aqueous solution. American Journal of Applied Sciences. 2016;**13**(6):814.826

[27] Zhang J. Phenol removal from water with potassium permanganate modified granular activated carbon. Journal of Environmental Protection. 2013;**4**(5):411-417. DOI: 10.4236/jep.2013.45049

[28] Kadam MM, Dhopte KB, Jha N, Gaikar VG, Nemade PR. Synthesis, characterization and application of c-MnO_2/graphene oxide for the selective aerobic oxidation of benzyl alcohols to corresponding carbonyl compounds. New Journal of Chemistry. 2016;**40**:1436

[29] Prasankumar T, Vigneshwaran J, Bagavathi M, Jose S. Expeditious and eco-friendly synthesis of spinel $LiMn_2O_4$ and its potential for fabrication of supercapacitors. Journal of Alloys and Compounds. 2020;**834**:155060

[30] Freitas Neto DB, Xavier FFS, Matsubara EY, Parmar R, Gunnella R, Rosolen JM. The role of nanoparticle concentration and CNT coating in high-performance polymer-free micro/nanostructured carbon nanotube-nanoparticle composite electrode for Li intercalation. Journal of Electroanalytical Chemistry. 2020;**858**:113826

[31] Molenda M, Dziembaj R, Piwowarska Z, Drozdek M. Electrochemical properties of C/$LiMn_2O_{4-y}S_y$ ($0<y<0.1$) composite cathode materials. Solid State Ionics. 2008;**179**:88-92

[32] Yue HJ, Huang XK, Lv DP, Yang Y. Hydrothermal synthesis of $LiMn_2O_4$/C composite as a cathode for rechargeable lithium-ion battery with excellent rate capability. Electrochimica Acta. 2009;**54**:5363-5367

[33] Yolshina VA, Yolshina LA, Elterman VA, Vovkotrub EG, Shatunova AA, Pryakhina VI, et al. Synthesis of and characterization of freestanding, high-hierarchically structured graphene-nanodiamond films. Materials and Design. 2017;**135**:343-352

[34] Yolshina LA, Yolshina VA, Pershina SV, Pryakhina VI. Study of thermal stability of hierarchical structured carbon composite flakes. Diamond and Related Materials. 2021;**119**:108556

[35] Yolshina VA, Yolshina LA, Pryakhina VI. SEM and XPS study of Cr^{6+} removal from wastewater via reduction and adsorption by hierarchically structured carbon composite in neutral media. Journal of Inorganic and Organometallic Polymers and Materials. 2021;**31**(8):3624-3635

[36] BTCORP. Nano-Materials For Sustainable Future. 2025. Available from: https://www.bt-corp.co/

[37] Moore M. Imaging diamond with x-rays. Journal of Physics: Condensed Matter. 2009;**21**:36421. (15pp). DOI: 10.1088/0953-8984/21/36/364217

[38] Prawer S, Nugent KW, Jameieson DN, Orwa JO, Bursill LA, Peng JL. The Raman spectrum of nanocrystalline diamond. Chemical Physics Letters. 2000;**332**:93-97

[39] Mehedi H, Arnault J-C, Eon D, Hébert C, Carole D, Omnes F, et al. Etching mechanism of diamond by Ni nanoparticles for fabrication of nanopores. Carbon. 2013;**59**:448-456

[40] Piazza F, Gough K, Monthioux M, Puech P, Gerber I, Wiens R, et al. Low temperature, pressureless sp^2 to sp^3 transformation of ultrathin crystalline carbon films. Carbon. 2019;**145**:10-22

[41] Melnik NN, Yu D, Korobov OSP, Savranskii VV. Raman spectra of carbyne-fullerene structures. Bulletin of the Lebedev Physics Institute. 2013;**40**(7):187-190. DOI: 10.3103/S1068335613070038

[42] Zeng Q, Wang L, Shi L, Qin K, Huang J, Tang K, et al. Optimizing

hydrogen plasma etching process of preferred (110)-textured diamond film. Surface & Coatings Technology. 2013;**228**:S379-S381

[43] Trócoli R, Morata A, Erinmwingbovo C, La Mantia F, Tarancón A. Self-discharge in Li-ion aqueous batteries: A case study on $LiMn_2O_4$. Electrochimica Acta. 2021;**373**:137847

[44] Taddesse P, Gebrekiros H, Semu G, Duressa M, Chemeda YC, Murali N, et al. Investigation of structural, vibrational spectroscopic and properties study of $LiMn_2O_4$ and LiMn1.9Cu0.05Fe0.05O4 cathode materials. Results in Materials. 2021;**12**:100224

[45] Biesinger MC, Payne BP, Grosvenor AP, Lau LWM, Gerson AR, Smart RSCC. Applied Surface Science. 2011;**257**:2717. DOI: 10.1016/j.apsusc.2010.10.051

[46] Dorogova VA, Elshina LA. Synthesis of hybrid carbon nanocomposites with potassium spinel KMn_2O_4 at low temperatures in a neutral medium. Russian Metallurgy (Metally). 2022;**8**:906-913

[47] Yolshina LA, Dorogova VA, Pryakhina VI. X-ray photoelectron spectroscopy study of the reduction and adsorption capacities of carbon nanomaterials using the formation of Mn(VII) ions. Russian Metallurgy (Metally). 2023;**8**:1112-1121

[48] Dorogova VA, Yolshina LA, Pryakhina VI, Vovkotrub EG. Reduction of MnO_4^- ions and selective deposition of sodium-manganese spinel nanocrystals on the surface of hierarchically structured carbon films in aqueous solutions. Journal of Inorganic and Organometallic Polymers and Materials. 2024;**34**(8):3527-3546. DOI: 10.1007/s10904-024-03017-3

Chapter 7

Heterogeneous Hydrogenation in Pharmaceutical Development

Alan Steven

Abstract

This chapter discusses an efficient means of robustly developing a heterogeneous hydrogenation for use in the development and manufacture of an active pharmaceutical ingredient, or an intermediate thereof. Reaction conversion data is used to establish a working expression for the differential rate that can be used to test a mechanistic hypothesis. Heuristics are shared that can be used when assessing how to keep the reaction under kinetic control. These can be used for problem diagnosis as well as scaleup strategy, and optimisation. A number of practical tips are shared, including the use of the process scheme as an information- and visually rich tool for presenting process understanding to a multidisciplinary team. Considerations when choosing the catalyst, solvent and equipment setup are also discussed.

Keywords: hydrogenation, heterogeneous, k_La, kinetic, trickle bed

1. Introduction

Hydrogenation is the addition of hydrogen across multiple bonds. Its facilitation by heterogeneous catalysts is a popular methodology for active pharmaceutical ingredient (API) synthesis. This discussion will include using hydrogen to cleave single bonds, as with debenzylations and *N*-benzyloxycarbonyl removals, transformations more correctly referred to as hydrogenolysis. In a review from 2006, the frequency of these reactions and nitroarene reductions, ensured heterogeneous hydrogenation was a prominent technology used for deprotections and reductions by the development and manufacturing groups of three large pharmaceutical companies [1]. In line with this popularity, between 10 to 20% of all chemical steps in the synthesis of APIs at a separate major pharmaceutical company have been estimated to be heterogeneous hydrogenations [2]. The complete menu of transformations of importance is available through the applications tables of some of those that make and sell heterogeneous catalysts [3, 4]. These companies include Chimet, BASF, Evonik-Degussa, Heraeus, Hindustan Platinum, Johnson Matthey and Strem.

The popularity of the technology can be attributed to the atom economy of the transformations, high yield and straightforward post-reaction processing. These characteristics address economic and throughput drivers that inform process design, as well as the need to control the amount of waste generated per kilogramme of processing stage product. Where the catalyst can be immobilised, its post-reaction removal is not required, simplifying process design still further.

The complexity generated by the presence of at least three phases, hydrogen gas, process solution and heterogeneous catalyst, has spawned multiple examples of scaleup issues [5]. The present work focuses on a development workflow and sharing know-how, with a view to achieving a "right first time" outcome. An envisaged use case is where a project wishes to scaledown a candidate process from the only available equipment train. Doing so allows understanding to be built around plant modifications required for scaleup, when it does takes place, to meet requirements around product quality, process safety, productivity and waste generation. Alternatively, the workflow could be used to characterise an existing heterogeneous hydrogenation process to allow an assessment of its "fit" with the various assets in a manufacturing network. Due to space constraints, transfer hydrogenation will not be covered. Catalyst-controlled examples of asymmetric induction with heterogeneous catalysts [6], a field dominated by homogeneous methods, will also not be covered.

2. Consideration for early process design

Once a project has agreed the transformation of a starting material A to a product AH_2 will feature in the route used to make an API, early Process Design activities will assess different ways of achieving the transformation. This includes establishing the reaction type (e.g. heterogeneous hydrogenation), categorical variables (e.g. solvent, and any reagent or catalyst) and the unit operations (e.g. charging, reaction, filtration, cleaning) involved. A degree of process parameter optimisation and characterisation is also performed, driven not least by the need, alongside the exploration of process knowledge space, to manufacture API for clinical trials. It is never too early to capture the 'musts', 'wants', 'avoids' and 'prefer nots' of heat and mass transfer, material balance, materials of construction, measurements required, mixing, operability constraints, physical processing requirements and residence time. Iteratively addressing gaps in knowledge in these areas, ensures decisions around duty design and equipment selection are data driven.

Many considerations for Early Process Design can be summarised in **Figure 1**. This Process Scheme includes the chemical elementary steps going on, and can be annotated with information on their kinetics, as well as relevant heat transfer equilibria, mass transfer, mass transport and phase dispersion data. Whilst simplistic in its format, and not to scale, it is visually rich and allows chemists and chemical engineers to understand the concerns and knowledge gaps of the other's discipline. In development, by assessing the effect of parameter changes on the chemical and mass transfer processes highlighted in the scheme, driving forces can be tentatively identified that enhance the rate forming the desired product and selectivity versus impurities. Experiments that fill in gaps in knowledge about these driving forces can then be judiciously designed, building up knowledge more rapidly than a statistical-based method for creating understanding.

An early activity is to establish the solubility of reaction components in solvents under consideration. The solvent needs to be able to dissolve the substrate A and hydrogen, for the reaction takes place. It should also keep the product AH_2 in solution, until any suspended catalyst has been filtered off and the composition adjusted to enable product crystallisation, whilst controlling the mass intensity of the entire process. The saturation concentration of hydrogen in organic solvents is only a few mmol/L and increases with temperature (except with predominantly aqueous systems) and headspace pressure [8]. The correlation with the latter is linear at low overpressures (Henry's Law).

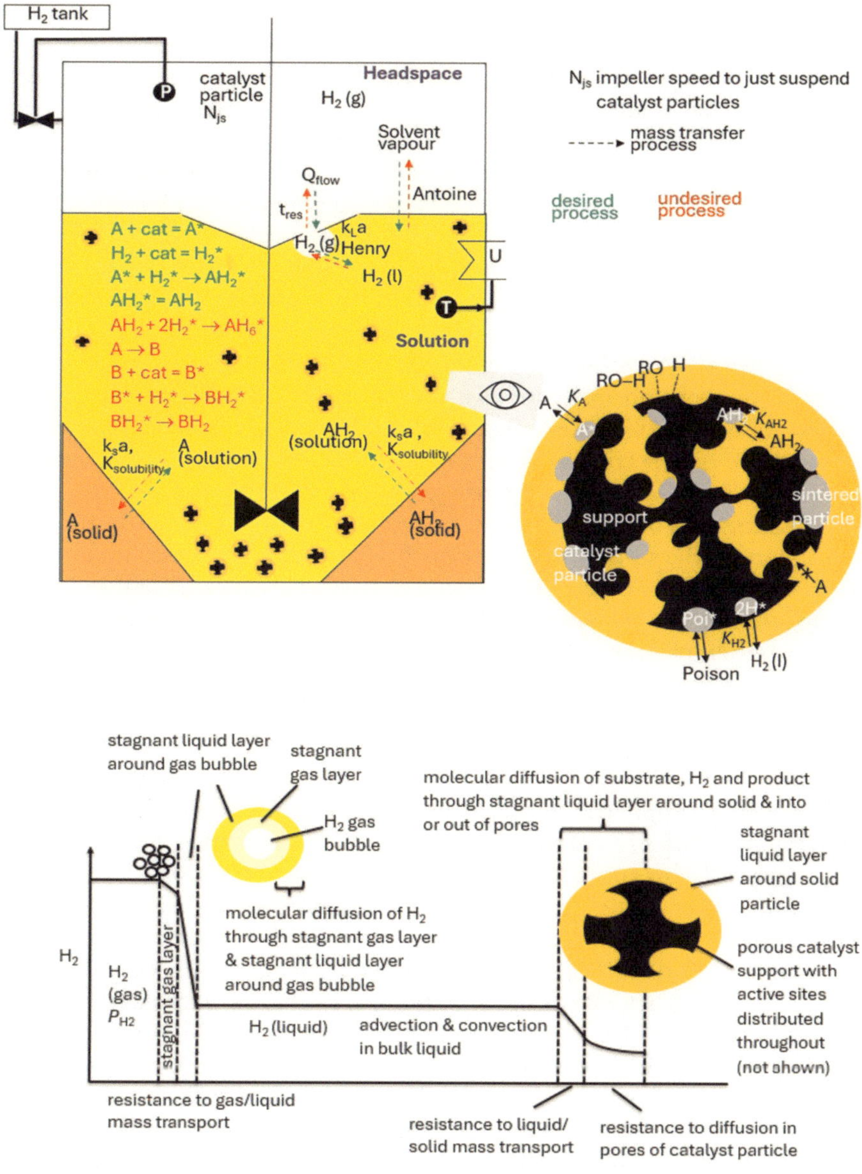

Figure 1.
Hydrogenation process scheme (top) and explanation of gradients in hydrogen concentration (bottom). Source: Adapted from Reference [7], https://www.mt.com/dam/non-indexed/po/autochem/clean/application-notes/hydrogenation-nitrobenzene-to-aniline/fundamentals_of_masstransferandkineticshydrogenation.pdf.

As a solvent, methanol should be chosen with caution as it can be oxidised to formaldehyde during heterogeneous hydrogenation catalysis. This leads to byproducts arising from its condensation with primary or secondary amines. Another caveat around the use of hydrogen with methanol or any other solvent with a low flash point is the risk of ignition. Procedures using aliphatic alcohols should also be wary of slurrying Raney nickel and supported palladium catalysts for an extended period of time, prior to the introduction of hydrogen gas. Under these circumstances, the catalyst surface is slowly poisoned through the generation and adsorption of carbon monoxide [9].

The use at elevated temperature and modest overpressure of a volatile material as solvent, generates appreciable solvent vapour pressure, at the expense of the partial pressure of the hydrogen in the headspace. On a related note, if not compensated for, the headspace partial pressures of carbon dioxide offgassed from an N-benzyloxycarbonyl removal will reduce hydrogen partial pressure.

The metal catalyst can either be supported or not. Raney nickel and Raney cobalt are unsupported metal catalysts with a high surface area whose formation are accompanied by the adsorption of hydrogen into the "sponge metal" surface. They remain workhorse materials in spite of requiring careful handling to address a pyrophoricity hazard. Their use also brings occupational exposure hazards. As well as being a possible human carcinogen and teratogen, when it is inhaled, nickel has been linked to pulmonary fibrosis and pneumonitis, whilst skin exposure causes skin rashes ("nickel itch").

Supported catalysts normally involve platinum group elements, with palladium offering the most versatility. The metal typically exists as ensembles hosted by an inert support like high surface area activated carbon, barium sulfate, calcium carbonate or a ceramic-based material. Regarding catalysts supported on carbon, powdered and paste catalysts are both used. Paste catalysts contain approximately 50–60% water within the pores of the carbon. Versus the use of powders, this attenuates both dusting and the risk of fire in the presence of vapours of organic solvent. Catalytic activity can be modified through the addition of modifiers [10]. Diffusion of reactants into the interior of a carbon particle potentially slows the reaction, so formulation for ("egg-shell") surface impregnation is the default. Increasing the saturation concentration of hydrogen in solution, forces hydrogen deeper into the pores, bringing more of any palladium distributed throughout the particle (as in **Figure 1**) into play.

It is not uncommon to simply screen catalysts during early Process Design where there is confidence supply can be maintained over the lifecycle of the API. That is to say, once you have found a catalyst that works, stick with that particular grade, and indeed lot number! This approach is derisked by laboratory testing ("use testing") of a new lot of catalyst for minimum activity requirements before the bulk catalyst is released for use on plant. An enhanced approach characterises related variants of the candidate catalyst, and builds understanding around material attributes tied to its activity, selectivity, deactivation and filtration performance. This paves the way for a meaningful specification based on control of these attributes. The characterisation is often best done in collaboration with academic partners, with the requisite expertise, and can involve measurements of or using Brunauer–Emmett–Teller (BET) surface area, X-ray diffraction, particle size distribution, electron microscopies (scanning, transmission and high resolution transmission), particle shape, gas pychnometry, geometric density and skeletal density [11]. In the case of palladium catalysts, X-ray photoelectron spectroscopy gives the ratio of palladium(0) to palladium(II) oxide.

Catalyst particles used for continuous processing with a trickle bed reactor have additional requirements over those used in batch [12]. Whilst this technology is routinely used to make bulk chemicals (≥1000 tonnes/year), the catalyst particles so used are too big for the reactors used to make the lower tonnages required of a pharmaceutical product. As well as particle size and distribution requirements (50–200 μm), candidate catalysts are shortlisted based on strict requirements around their mechanical strengths (>1000 bar) and surface roughness (which relates to how easily metal can be stripped from the support surface), before being assessed for activity and selectivity. Raney catalysts are fine powders but they have still managed to be accommodated in trickle bed reactors for the hydrogenation of sub-kg quantities of API intermediates [13, 14].

Knowledge of the rate of degradation is an important development objective. It allows the calculation of the maximum amount of substrate that can be converted per unit mass of catalyst, by the time it has completely deactivated. Its establishment is particularly important in an immobilised setting, where there are expectations around higher utilisation versus batch. As well as poisoning by contaminants in the inputs (*vide infra*), attrition and metal stripping, thermal stress, sintering or solid-solid chemical reactions can additionally degrade the material [15]. Susceptibility to degradation is assessed by monitoring the stability of the reaction conversion over an extended period in a packed bed setting. The catalyst loading and residence time should target a starting point of ca. 80% (i.e. incomplete reaction). Where the catalyst is a powder, to be used in batch, it can be admixed with silicon carbide particles or glass beads in the packed bed setup for this testing.

Filtering off a heterogeneous catalyst at the end of a hydrogenation is attractive versus the metal remediation procedures of homogeneous catalysis, but adsorbed hydrogen will remain on the catalyst surface, rendering the material pyrophoric and a fire hazard. Postreaction, Raney catalysts can be deactivated by washing with sodium nitrite in water [16]. According to the Good Manufacturing Practices that underpin the latter stages of API manufacture, spent catalysts are not normally recycled as is, but the material can be returned to the catalyst manufacturer for metal recovery after it has been put under water. Where catalyst recycling is permitted, it can be returned to its makeup vessel, as a slurry, once it has been pushed off the filter medium,

3. Characterising the intrinsic kinetics of the reaction

For a heterogeneous hydrogenation to be scalable and portable, the rate of the chemical reaction at the catalyst surface should be so slow, the observed (extrinsic) rate is not limited by scale- and equipment-dependent phenomena, namely mass transfer, mass transport and heat transfer. The intrinsic kinetics at the catalyst surface depend on the equivalent of catalyst concentration, substrate concentration, temperature and the concentration of hydrogen in solution. The full rate expression is a characteristic of the reaction, over the parameter ranges where the associated mechanism applies, and is independent of equipment.

It is critical that hydrogenation development builds up working knowledge of the rate expression, commensurate with the phase of development of the product. Time course data for the amount of hydrogen taken up by the system at constant pressure can be used to make inferences about the mechanism, and examples are given in **Figure 2**. These data should always be recorded with a calibrated flowmeter. Where there are mass balance issues, due to inadequate calibration, these can be corrected for by ratioing the hydrogen actually consumed by the sum of that required by the chemistry and that needed to saturate the solvent. There may also need to be correction for any ongoing leakage from the reactor. An orthogonal but still "global" signature is the heat outputted as a measure of time during reaction calorimetry [17]. This heat flow varies directly with the reaction rate. Neither hydrogen uptake nor heat evolution differentiate between the desired and undesired chemical processes.

FT-IR and in situ Raman are good for monitoring individual species, including unstable intermediates, but are not sufficiently sensitive to accurately measure low levels of impurities. They can be used at low overpressures and do not require method development. Offline chromatographic measurements offer specificity and sensitivity that allow starting material, product, stable intermediates and impurities to all be

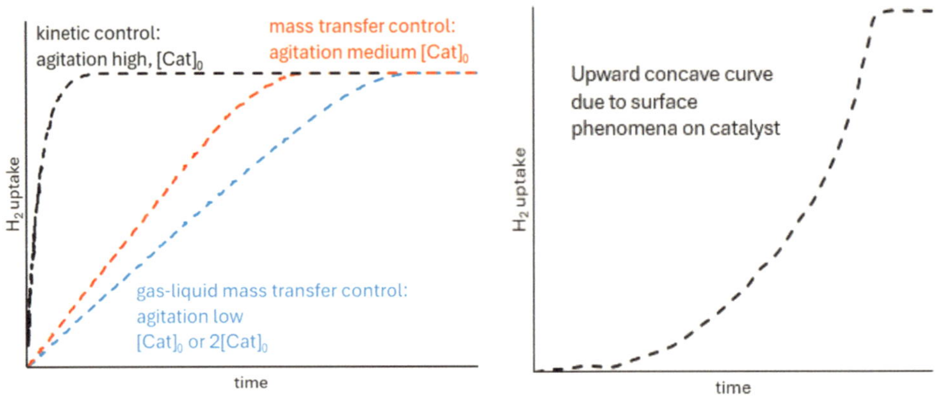

Figure 2.
Examples of hydrogen uptake graphs.

quantified, if the setup is able to sample under pressure and there is a suitable chromatographic method. The most appropriate measurement technology will depend on the needs of development, but studies should strive to measure hydrogen uptake as a minimum. When more than one elementary step is involved, the determination of kinetic parameters becomes increasingly accurate as more intermediate species and impurities are quantified, along with the starting material and product.

Table 1 presents a typical workflow to generate data for making kinetic inferences. The baseline temperature should be the temperature anticipated for the final process. The reactions should be performed isothermally, using a low substrate concentration and a low catalyst loading to control any exotherm, and with good gas-liquid mass transfer. A rigorous way of confirming good gas-liquid mass transfer, in this or any other study, is to change the gas-liquid mass transfer coefficient (*vide infra*) to establish no effect on the overall hydrogenation rate (compare with **Figure 2**, left). The same batch of catalyst should be used across the experiments to maintain the number of mol active sites/g of catalyst.

$$\mathrm{A} + \mathrm{H_2} + \mathrm{Cat} \rightarrow \mathrm{AH_2} + \mathrm{Cat} \qquad \frac{-\mathrm{d}[\mathrm{A}]}{dt} \approx k[\mathrm{Cat}][A][H_2]_{sat} \tag{1}$$

When choosing a candidate mechanism, whose corresponding rate expression can be assessed against observed data, it is useful to start off as simple as can be. In the absence of any further knowledge about the mechanism, a useful starting point for a well suspended catalyst is to effectively treat it like a dissolved reagent, as in Eq. 1, where it is assumed the conditions keep the solution saturated with hydrogen. This is first order in all of the inputs. The expression often fits the experimental results, but has little to say about the actual surface reaction mechanism of the process. Fitting a working candidate rate expression against the data from runs 3–13 in **Table 1** will generate values for missing mechanistic parameters. This is the rate constant k in Eq. 1, and will additionally include adsorption constants in more complex cases, and activation energies where the temperature is varied. Where intermediates in a multistep sequence are not detected during reaction monitoring, despite the analytical method being capable of their detection, they are not likely to feature in the turnover-determining step. The rate constants associated with their formation can be removed from the fitting exercise by fixing them at arbitrary large values which will not affect

Run	[Substrate]	[Catalyst]	P_{H_2}	Temperature	Stirring	
1	b'line	b'line *5	0.5 barg	b'line	slow	a
2	b'line	b'line	b'line	b'line	normal	b
3	b'line	b'line	b'line	b'line	normal	c
4	b'line	b'line	b'line	b'line	normal	d
5	b'line	b'line	b'line	b'line	normal	
6	b'line	b'line	b'line	b'line	fast	d, e
7	b'line	b'line *2	b'line	b'line	normal	e
8	b'line	b'line	b'line /2	b'line	normal	
9	b'line /2	b'line	b'line	b'line	normal	
10	b'line	b'line /2	b'line	b'line	normal	
11	b'line	b'line /4	b'line	b'line	normal	
12	b'line	b'line	b'line *2	b'line	normal	
13	b'line *2	b'line	b'line	b'line	normal	
Fit model against runs 3–9 (lean) or 3–13 (full)						
14	b'line/2 < x < b'line	b'line/2 < x < b'line	b'line/2 < x < b'line	b'line	normal	f
Fit model against runs 3–9 + 14 (lean) or 3–14 (full)						
15	b'line	b'line	b'line	b'line + 10 °C	normal	
16	b'line	b'line	b'line	b'line - 10 °C	normal	

P_{H_2} Hydrogen pressure *b'line baseline condition.*
a) Hydrogen "starvation" run to accentuate side reactions not involving hydrogen.
b) Delay hydrogen admission by 2 hours, versus run 4, to accentuate any background reaction, including inhibition of catalyst by other reaction components.
c) Product inhibition run: Prestir catalyst slurry with a stoichiometric amount of AH_2 prior to the introduction of the substrate.
d) Runs to establish reaction reproducibility using baseline parameter values.
Lower the catalyst loading if the reaction time is less than 10 min.
Analyse an End of Reaction sample for leached metal, after filtration.
Evaluate how $\frac{[product]}{[byproduct]}$ ratio changes with time, including on holding well beyond the point of complete substrate consumption.
e) Used to establish reaction is in the kinetic regime, in which case set $k_L a > 10/s$ in the absence of a more accurate estimate. Also used to pinpoint any source of mass transfer resistance when the reaction is not under kinetic control.
f) Validation run.

Table 1.
Workflow to generate data for kinetic modelling

the overall rate. Fitting involves (simultaneously) changing the parameter(s), using an algorithm like the Levenberg-Marquardt method, until the sum of squares of the residuals between the curve for the simulated data and the experimental data has been minimised across the different experimental scenarios being assessed. These studies require operation with good gas-liquid mass transfer.

Where there is a good fit between the curves for the simulated data and the experimental data across the different experimental scenarios assessed, a further validation run (run 14 in **Table 1**) is run to assess the predictive capability of the model. This validation run should use previously untested values of [Catalyst], [Substrate] and P_{H_2}. Assuming the model's predictions are reasonably accurate, the model can be revalidated to include this extra data set. Activation energies can be fitted after generating data at different reaction temperatures (runs 15 and 16 in **Table 1**, also see **Figure 3**, bottom).

Fitting of the initial data sets and attempts to validate the model, by testing its predictions, are often clearly unsuccessful. In this case, where no reasonable experimental error can be assigned, the elementary steps behind the mechanism, and the assumptions about their kinetic relevance, will need to be updated. This means building complexity into the simplified semiempirical starting point (Eq. 1), and considering the kinetic relevance of the processes happening on the binding surface of the catalyst, before revisiting the modelling exercise. In this regard, a mechanistic model commonly implicated next is the so called Langmuir-Hinshelwood-Hougen-Watson (LHHW) dual site mechanism [18]. Substrate A, product AH_2 and hydrogen compete with one another for adsorption onto a finite number of active sites present on the surface of a catalyst, at rates proportional to both the solution concentrations of the species and the number of uncovered active sites on the catalyst. If the substrate is preferentially adsorbed, the degree of adsorption of the hydrogen will drop, and vice-versa. Once adsorbed onto the catalyst surface, the species react to form adsorbed

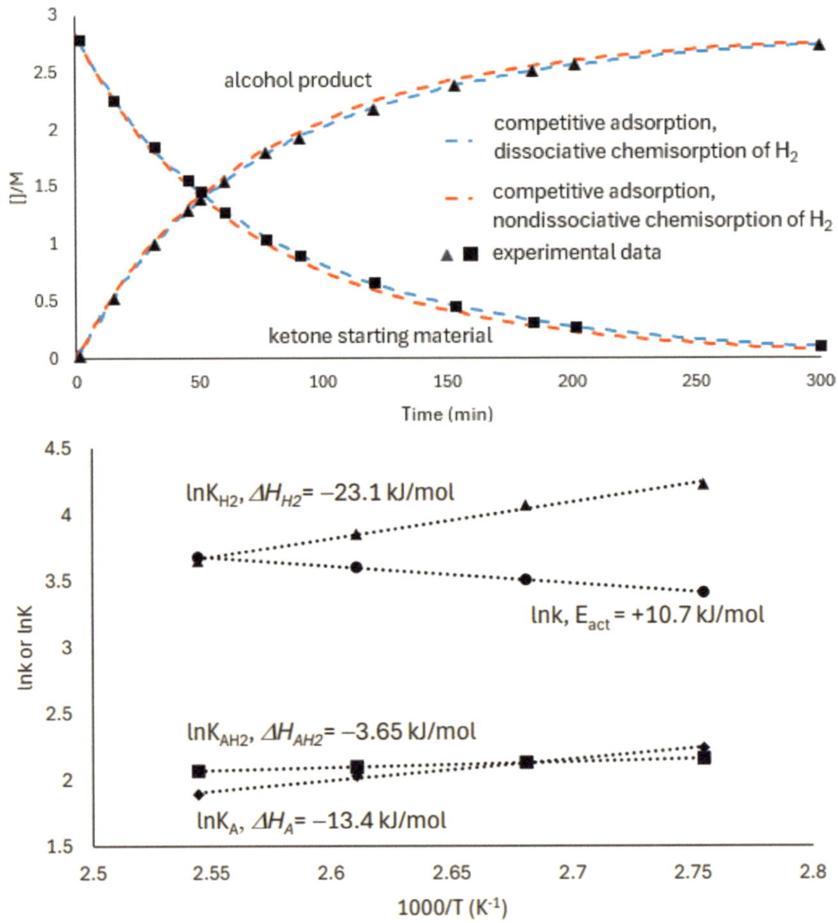

Figure 3.
Fitting of time course data against different mechanistic proposals (top) for the reduction of a proprietary ketone and extraction of temperature-related parameters (bottom). Source: Data provided by author.

product at a rate proportional to their respective concentrations on the catalyst surface. Intuitively, the probability of two adsorbed species colliding is low, and indeed this mechanism relies on this reaction being turnover-limiting. Depending on its affinity for the catalyst, product AH_2 can compete with the substrate A and H_2 for catalytic sites, such that the desorption of AH_2 into the solution phase may be appreciably reversible. The associated rate expression is shown by Eq. 2. Clues as to the need to apply this model are any of the following:

- The reaction is second order with respect to the catalyst concentration
- Hydrogen uptake with time shows a concave profile (where the adsorption of hydrogen is slow relative to substrate A) (**Figure 2**, right).
- A reaction is faster at lower agitator speed consistent with the slow adsorption of A relative to the adsorption of H_2.
- The reaction rate decreases with increased hydrogen pressure, if the substrate A is less effective at competing with hydrogen for active sites.

$$\begin{array}{ll} \mathrm{A(l)} + \mathrm{Cat(s)} \rightleftharpoons \mathrm{A}^* & K_A \\ \mathrm{H_2(l)} + \mathrm{Cat(s)} \rightleftharpoons \mathrm{H_2^*} & K_{H_2} \\ \mathrm{A}^* + \mathrm{H_2^*} \rightarrow \mathrm{AH_2^*} & k_3 \\ \mathrm{AH_2^*} \rightleftharpoons \mathrm{AH_2\ (l)} & K_{AH_2} \end{array}$$

$$-\frac{d[A]}{dt} = k_3[Cat]_{total}^2\theta_A\theta_{H_2} = k_3[Cat]_{total}^2\frac{K_AK_{H_2}[A][H_2]}{(1+K_A[A]+K_{H_2}[H_2]+K_{AH_2}[AH_2])^2} \quad (2)$$

where $[Cat]_{total}$ is the total molar amount of catalyst (free plus that bearing adsorbate) and $\theta_X = \frac{K_X[X]}{(1+K_A[A]+K_{H_2}[H_2]+K_{AH_2}[AH_2])^2}$, the expanded Langmuir isotherm

Where $K_A[A]$ dominates the denominator in Eq. 2, as with a tightly bound substrate, the reaction order with respect to the substrate A is -1 over much of its course, and its fed batch addition should be considered to address its inhibitory effect. Where $K_{H_2}[H_2]$ dominates the denominator, as with a weakly bound substrate, the rate varies as inverse pressure. When $K_A[A], K_{H_2}[H_2]$ and $K_{AH_2}[AH_2]$ are all small compared to 1, the expression simplifies to $\approx k_3[Cat]_{total}^2K_A\ K_{H_2}[A][H_2]$. The practical relevance of $K_{AH_2}[AH_2]$ to the denominator, and hence the inhibitory effect of AH_2, can be established by run 3 in **Table 1**. When the above mechanism is being tested, the rate constants for the forward reactions of the adsorption equilibria can be fixed at arbitrary high values where solid-liquid mass transfer is known not to be rate-limiting, such that concentrations of catalyst-bound species quickly equilibrate. This simplifies the fitting problem by leaving only the equilibrium constants for adsorption between solution and catalyst-bound forms of A, H_2 and AH_2 to be fitted, alongside activation energies and the rate constant for the reaction between catalyst bound species. As the LHHW model is truly mechanistic, unlike Eq. 1, it can be cautiously extrapolated to make predictions about performance outside of the parameter ranges covered by the data used for the original fitting.

Eq. 2 relates to a scenario where the chemisorbed hydrogen appears to react as a molecular adsorbate. In reality, the chemisorbed hydrogen dissociates into atoms that diffuse across the catalyst surface ahead of stepwise addition to A* [19]. The addition of the first hydrogen atom is reversible, and bond rotation leads to the potential observation of the geometric isomer of a diastereochemically pure alkene input, over the course of any alkene hydrogenation. For the reduction of a proprietary pharmaceutical intermediate shown in **Figure 3** (top), accounting for the dissociative chemisorption of hydrogen, such that the square root of the hydrogen concentration features in the differential rate expression, amongst other changes [20], produced a better fit than Eq. 2 (nondissociative chemisorption). Repeating the experiment at different temperatures and plotting the fitted values of the rate constant and the adsorption constants against inverse temperature allow the extraction of the activation energy and enthalpies of adsorption (**Figure 3**, bottom). This involves multiplying the gradients of the lines of best fit (units K) by R = −8.314 J/(mol•K).

A small numbers of cases, where the observed kinetics indicate only strong adsorption of A and not H_2, are often explained using the Langmuir-Rideal (single site) mechanism [21]. Whilst not likely to be a true reflection of reality, as the adsorption of hydrogen is largely considered mandatory [22], it has been traditionally formulated as the mechanism alongside Eq. 3, with A* apparently reacting with non-adsorbed H_2 in the slow step. Applying the steady state approximation to the concentration of the adsorbed complex and again assuming it is the chemical reaction which is turnover-limiting, the rate is as shown in Eq. 3. The rate depends on [A*], which is largely a constant ([Cat]), given the strong binding of A. The rate is thus pseudo-zero order in A for the majority of the reaction's course, reverting to first order at the very end.

$$\begin{aligned} &\mathrm{A}(l) + \mathrm{Cat(s)} \rightleftharpoons \mathrm{A}^* && K_A && -\frac{d[A]}{dt} = k_2[Cat][H_2]\,\frac{K_A[\mathrm{A}]}{1+K_A[\mathrm{A}]} \\ &\mathrm{A}^* + \mathrm{H_2(l)} \rightarrow \mathrm{AH}_2^* && k_2 && \\ &\mathrm{AH}_2^* \rightarrow \mathrm{AH_2}\ (l) && && \end{aligned} \quad (3)$$

The presence of poisons in the inputs that quickly and irreversibly deactivate the catalyst can be established by plotting the initial rate versus the catalyst loading (runs 4, 7, 10 and 11, **Table 1**), when input materials representative of those to be used in the manufacture become available. Any non-zero intercept where the extrapolated form of the line of best fit hits the catalyst loading axis is the poisoning threshold, an amount of catalyst that can effectively be written off. A poison is sometimes a high molecular weight oligomeric contaminant in the substrate, and difficult to pick up without size exclusion chromatographic analysis. Where this contaminant cannot be removed, a control strategy is to employ a catalyst whose metal is distributed throughout a carbon support, such that size exclusion by the support's pores keeps the contaminant away from the bulk of the metal. Other common poisons are carbon monoxide in hydrogen and sulfur compounds in the substrate. Despite attempts to exclude poisons from the input materials, they may still arise due to carryover from the previous process to be used in the manufacturing equipment. This can be picked up by introducing the catalyst to the equipment during a solvent trial used to establish the behaviour of unit operations like distillations, before they are executed for real. The catalyst used in this sacrificial run is filtered off and tested in the lab. If the recovered catalyst shows no activity, further sacrificial runs should be repeated, and the actual manufacture paused until the recovered catalyst shows activity in the lab.

4. Hydrogenation mass transfer

$$\frac{-1}{V_{rxn}}\frac{dmol_{H2,g}}{dt} = \frac{d[H_2]_{sol}}{dt} = k_La\left([H_2]_{sat} - [H_2]_{sol}\right) - n_{H_2}r_{int} \tag{4}$$

where n_{H_2} is the number of equivalents of hydrogen used to reduce A to AH_2 and r_{int} is the rate of the chemical reaction.

The hydrogen in solution that feeds the reaction on the catalyst surface needs to be replenished by the dissolution of headspace hydrogen. These processes are captured in Eq. 4, an expression for the rate of change of concentration of hydrogen in solution. The rate of dissolution depends on k_La, the volumetric gas-liquid mass transfer coefficient and the thermodynamic driving force. The latter is the difference between the saturation concentration $([H_2]_{sat})$, as predicted by Henry's Law at low overpressures, and the actual solution concentration $([H_2]_{sol})$. If the chemical reaction is much slower than hydrogen dissolution (a requirement for its study, see previous section), the first part of the right-hand side of Eq. 4 disappears and the rate of hydrogen uptake approximates to the rate of the chemical reaction. If the chemical reaction is much faster than hydrogen dissolution, the overall rate is limited by the movement of the hydrogen from the headspace to point where it can chemisorb on the hydrogen-starved catalyst surface. In the extreme case, $[H_2]_{sol}{\approx}0$, and the rate of change of the solution concentration of hydrogen simplifies to its maximum value, $k_La[H_2]_{sat}$ = k_La $H_{T,solvent}P_{H_2}$, where $H_{T,solvent}$ is the Henry constant and P_{H_2} is the pressure of hydrogen. This ensures time course data for the transformation is zero order in substrate until the chemical reaction slows down.

The most practical lever that can address this mass transfer limitation is k_La, as the hydrogen pressure is typically limited by supply and vessel design limits. This coefficient varies substantially between vessel designs, as it depends on reactor geometry, agitation rate, impeller type, impeller location, any baffling and any subsurface sparge line. Even small changes to the setup of a reactor can have an unexpected effect on the rate, where the reaction is under hydrogen mass transfer control. The effect is disproportionate with changes in agitation rate as it is common for k_La to exponentially depend on this parameter. Whilst relating to a lab study, rather than a change of vessel or facility, **Figure 4** illustrates how a change in $[H_2]_{sol}$, exercised using agitation rate to change k_La, can dramatically affect an outcome (in this case reaction enantioselectivity) [23].

Where a technology transfer has ignored the consideration of k_La, unless the reaction at the catalyst is sufficiently slow, maintaining the same headspace pressure (as in **Figure 4**) will not be sufficient to ensure the relative rate of competing processes is maintained across the two vessels. Alongside this consideration is the prolonged reaction time associated when gas-liquid mass transfer replaces the intrinsic reaction as the bottleneck for the observed rate. This leads to higher levels of background reactions that do not rely on the catalyst. Deactivation of the catalyst due

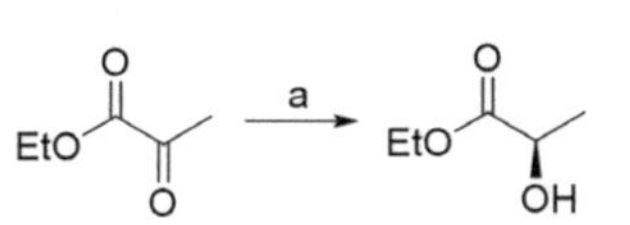

2000 rpm, ee=60%; 400 rpm, ee=23%

a) cinchonidine-modified Pt, 30 °C, H_2 (70 psig), *i*-PrOH

Figure 4.
Example of the dependence of ee on k_La.

to its offcycle poisoning or undersaturation with hydrogen can also potentially become more prevalent.

Given the importance of knowing when a hydrogenation is likely to be at risk of gas-liquid mass transfer control, attempts should be made to measure k_La for all hydrogenators used in the lifecycle of a hydrogenation reaction. Where k_La cannot be measured, it is modelled based on reactor geometry, configuration and fluid properties [24]. The solvent used for the reaction medium affects the fluid properties, and the coefficient also depends on reactor fill volume and temperature. Where accurate and fast pressure readings are possible, a method for the measurement of k_La for a reactor in a particular configuration uses hydrogen uptake data collected at various agitation speeds and vessel fills [25]. With the agitator off, and in the presence of a degassed solution that does not contain the reaction substrate, the headspace is pressurised with hydrogen gas before the hydrogen line valve to the hydrogenator is closed. The hydrogen feedline should preferably be preheated so the temperature in the headspace is close to that of the batch. Agitation is then started and the drop in the headspace pressure measured as a function of time. This is illustrated in **Figure 5**. The actual drop is related to the solubility of the gas in the liquid at that particular temperature but k_La is the slope of the plot versus time of the expression on the second y-axis of **Figure 5**.

When initially charged to the vessel from their storage container, the particles will contain catalytically inactive palladium hydroxide or surface oxidised palladium. The above experiment can be repeated in the presence of a supported palladium catalyst, but still without substrate. The difference in the uptake profiles, with and without catalyst, gives information on the kinetics of the reduction of catalyst particles to the catalytically active palladium(0) state.

An alternative approach to determining k_La is to use the hydrogen uptake of a test reaction performed in the mass transfer regime. The latter can be achieved by using a sufficiently high catalyst loading that the rate of hydrogen uptake is initially constant. Dividing this rate by $[H_2]_{sat} = H_{T,solvent}P_{H_2}$ gives k_La. To cut down on the effort involved in characterising k_La, it has been modelled using a statistical design of experiments approach using agitation speed, fill volume and temperature.

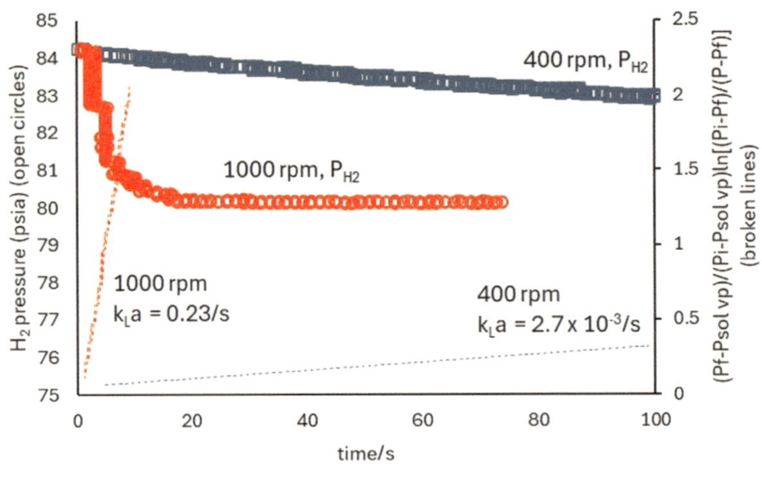

Figure 5.
Gas uptake for the calculation of k_La. *Calculation of* k_La *for hydrogen uptake by 1-propanol (0.5 L, 30°C).* P_f *= final pressure,* P_i *= initial pressure,* $P_{sol\ vp}$ *is the solvent vapour pressure. Source: Adapted from data in Ref. [26], Chemical Engineering in the Pharmaceutical Industry.*

Such a study should additionally look at the effect of solvent, or a proxy thereof such as the viscosity of a blend of solvents. Viscosity is a continuous variable and can be varied by tuning the composition of the blend.

Gas-liquid mass transfer can be a focus for troubleshooting given its dependence on scale and equipment, though it is by no means the only rate process involved in ensuring the catalyst is supplied with adequate amounts of hydrogen. The reciprocal of each of these rate processes can be thought of a resistance which acts as a drag on the observed rate of reaction (**Figure 6**). A second significant source of resistance is solid-liquid mass transfer. Where the substrate dissolves over the course of the reaction, as in the Process Scheme (**Figure 1**), the particle size affects the rate of dissolution. For slow-to-dissolve substrate and the heterogeneous catalyst, there is the need to adequately suspend particles of both to ensure the full surface area of these materials is available for dissolution and catalysing the reaction respectively. As **Figure 1** shows, solid-liquid mass transfer resistances at the catalyst particle involve both diffusion across the thin film separating the catalyst surface from the bulk solution and into the catalyst pores. These processes are dominated by the size and shape of the catalyst particles themselves, as well as the identity of the catalyst support. Thus, where the same batch of catalyst is being used for manufacture and development, even if solid-liquid mass transfer is limiting, performance should translate if it is well suspended. Where a reaction presents as zero order in substrate, and solid-liquid mass transfer at the catalyst is suspected as being limiting, deliberately increasing the number of active sites can be used to pinpoint the source of the resistance. If doubling the catalyst concentration (runs 4 and 7, **Table 1** and **Figure 2**, left) has a positive effect on the observed rate whilst retaining zero order behaviour with respect to substrate, solid-liquid mass transfer is important at the lower catalyst loading.

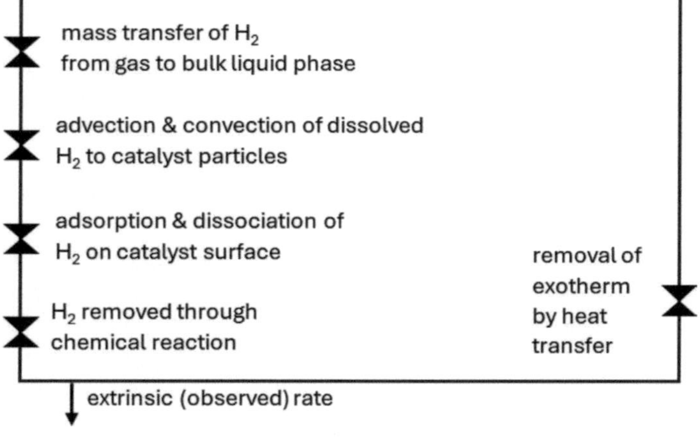

Figure 6.
Rate processes affecting the observed rate of a hydrogenation. Source: Adapted from Reference [5], https://dcresources.scale-up.com/Open.aspx?file=280&lid=281&c=sus.

The efficiency of heat removal from batch reactors used for slurry hydrogenations depends on the surface area of the cooling jacket and decreases in relative terms as the volume of the reactor increases. It is crucial that the enthalpy of the hydrogenation is considered early on as part of the process design for highly exothermic transformations like the reduction of a nitroarene to the corresponding aniline (−554 kJ/(mol nitrobenzene) in the liquid phase). This enthalpy, which also allows evaluation of the adiabatic temperature rise, is a useful byproduct of reaction monitoring by reaction calorimetry. Where it cannot be easily measured, it can be estimated by subtracting the cumulative bond energies of the products from those of the reactants [8]. The headspace hydrogen pressure can either be held constant or the supply switched on and off, depending on if a low pressure and high pressure are respectively present in the headspace. Either way, the setup should allow the replenishment of consumed hydrogen in a timely fashion. Otherwise, even when gas-liquid mass transfer is good, fast reactions run the risk of depleting the amount of hydrogen in the headspace to the point where the catalyst starves.

Most resistances operate in series but heat transfer operates in parallel (**Figure 6**). When a reaction starts off as not being under kinetic control, and reaction parameters are held fixed, the different resistances in **Figure 6** may take their turn in exerting primacy over the extrinsic rate. As each may well affect competing reactive chemistry pathways to different extents, the reaction selectivity, as well as being different from reaction in the kinetic regime, will change over the course of the reaction. We will now show how a quantitative treatment of these resistances can be considered alongside the model for the intrinsic rate, when accommodating a hydrogenation into a manufacturing facility.

5. Considerations when fitting a process into a facility

Figure 7 uses a number of heuristics to advise when any one of the sources of resistance discussed previously is likely to bottleneck the observed rate. These heuristics are discussed on the DynoChem Resources website [5]. Whilst the consideration of the different resistances is presented in a certain sequence here, they should be considered in an order that best suits the project. This workflow can be used during process deliberations around which hydrogenation facility offers the best "fit" for the reaction's kinetics. Alternatively, it can focus attention on improving the suitability of the sole site under consideration, for example, by highlighting the need to invest in improved mixing capability. To avoid the possibility of a runaway situation, the operating temperature and catalyst loading should be chosen so the heat generated does not overwhelm the heat transfer capability available.

Whilst the **Figure 7** workflow is powered towards keeping the reaction in the kinetically-controlled regime, there may be good reasons to purposefully keep a reaction in the mass transfer controlled regime. A reaction model built using knowledge of the relationship between k_La, the reaction rate and the amount of heat generated allowed the temperature of a highly exothermic nitroarene addition to be controlled by controlling the rate of agitation [27]. To ensure performance was maintained between lab and plant, k_La was maintained across both settings. This approach brings with it the risk of some irreversible deactivation of the undersaturated catalyst.

The enhanced heat transfer capability of a continuous processing technology would have been another possible solution for the above scenario. A modified

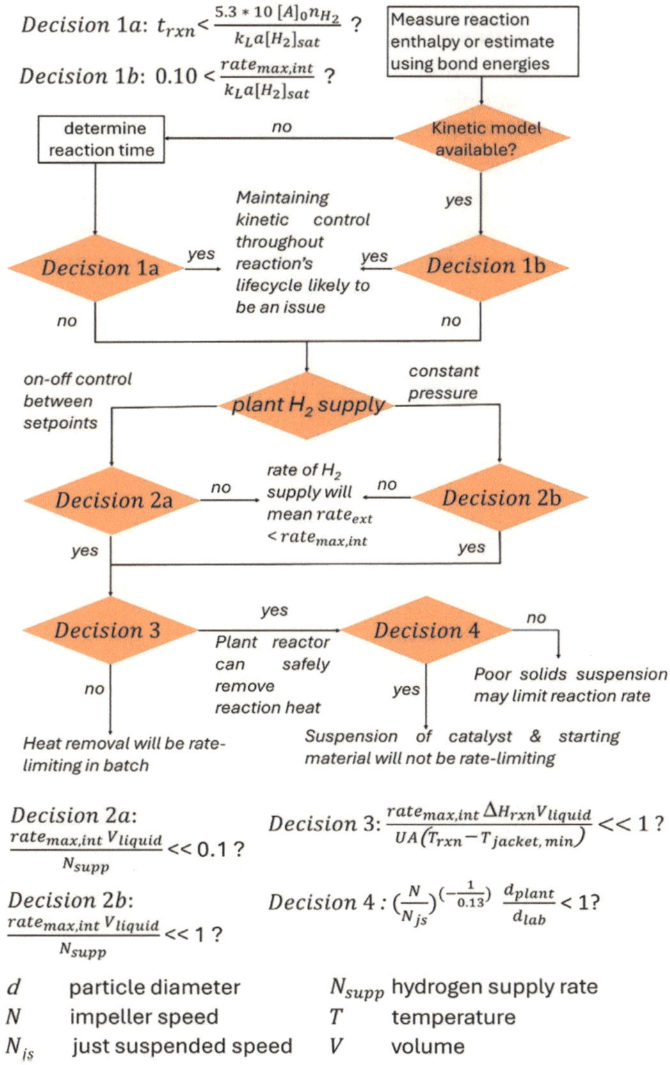

Figure 7.
Heuristics to assess process fit. Source: Adapted from data in Ref. [5], https://dcresources.scale-up.com/Open.aspx?file=280&lid=281&c=sus.

workflow is required for scaling up the use of a flow reactor using an immobilised catalyst bed. Catalyst suspension is not required here. The diameter of the reactor is increased whilst maintaining the height of the bed and the velocity of the process stream. This largely maintains the pressure drop at the larger scale, as modelled using the Ergun equation [28], mitigating the risk of a portion of the catalyst bed becoming undersaturated with hydrogen and deactivating.

Towards the end of Process Design, the project will have built out its kinetic model into a physical model with additional terms that are independent of scale and equipment. Characterisation ranges around operating setpoints are then agreed for parameters in the model. The model is virtually run thousands of times, as a full factorial DoE, to predict the process performance throughout the multidimensional hyperspace defined by these ranges. The part of the hyperspace that meets the targets for product quality (e.g. impurity formation), consistency and process performance (e.g. yield, hydrogen uptake rate, reaction rate) define a design space-like region that can be registered with Health Authorities as part of the process for gaining a licence to make the API for commercial use. A recent example is from AbbVie, where control was exercised over the levels of genotoxic intermediates in the product of a nitroarene reduction [29]. A means of simultaneously assessing ranges for multiple process parameters when establishing the shape of the design space-like region is to use a parallel coordinates plot.

Errors in the measured values used for model training, errors in 'noise' variables, general model error, and errors in the parameters fitted during modelling all produce variability in the actual response generated if any one virtual DoE experiment were to be repeatedly run in the lab. The k_La data is likely to present a particularly important source of uncertainty, particularly where it has been modelled (rather than measured) based on the reactor geometry. The cumulative variation for the different errors in the model can be accommodated using a Bayesian-based statistical approach.

The manufacturing partner ought to have a support lab with a miniaturised version of the plant reactor, which mimics the geometry and heat and mass transfer behaviour of the latter. This "twin" reactor is used for experimentation to establish that the ranges predicted by the virtual DoE are in fact suitable for use in the manufacturing environment envisaged for the commercial supply of the API. By experimenting close to the edge of acceptability of the design space-like region, this testing seeks to establish that even "worst case" combinations of inputs produce suitable material.

6. Further equipment considerations

We will close out the discussion by providing broader context around equipment considerations. Hydrogenations have mostly been carried out in stirred tank hydrogenation reactors [30]. Roessler has described a detailed process flow covering charging, inertisation, reaction, sampling, slurry filtration, waste treatment and safety engineering controls to address the process safety challenges of adding, removing and handling catalysts in slurry reactors [31]. The pressure in vessels used for pilot or production scale production will typically be limited to 5 bar on safety grounds, though some facilities have access to autoclaves that go up to 100 bar. In addition to a jacket for heat transfer, and a hydrogen supply, a conventional setup in such a reactor will have a pressure controller and a calibrated flowmeter to measure hydrogen uptake. This setup is frequently available in plant hydrogenators, and works well when all the solvents and feeds are batch charged to the reactor at the start. Catalyst splashed onto the walls above the tide mark of the reactor can be baked on during heating. The removal of this material requires both labour- and time-intensive manual intervention by the equipment operator, to the detriment of the length of a manufacturing campaign.

The use of pressurised hydrogen used by a stirred tank reactor represents a significant safety hazard and requires its own zone in a manufacturing plant. The higher the gas-liquid mass transfer (typical values <0.005–0.5/s), the greater the opportunity to use hydrogen in nearly stoichiometric amounts. To help with initial saturation of the batch, a gas-inducing agitator will supply hydrogen via nozzles at the impeller tips after it has been pulled down through the hollow shaft. Alternatively, the hydrogen can be supplied with a sparge ring located beneath the impeller, though this requires special blades that can transfer their power to the liquid, whilst the impeller is being flooded with hydrogen gas. With a normal stirred tank reactor operated in semibatch mode, a substantial change in the batch volume means the impellers and heat transfer surfaces are not properly submerged initially, compromising performance and giving rise to significant temperature anisotropy. The heat transfer performance of a standard stirred tank hydrogenator can be improved by fitting a loop to pump the reaction mixture through an external heat exchanger. This configuration allows the heat exchanger and pump to be sized separately from the reactor agitation system.

A stirred tank reactor with high heat transfer capability in and of itself is the Biazzi reactor. This uses highly efficient curved plate heat exchangers that are readily cleaned. The jet loop reactor is a specialist batch reactor which is not agitated and uses a gas-liquid venturi ejector to pull hydrogen into the reaction mixture (producing $k_La s > 10/s$). The jet loop reactor also uses an external heat exchanger. It works well if a solution of substrate is added over the course of a reaction, making it a useful asset for highly exothermic hydrogenation reactions. A variant of the jet loop reactor has a monolith catalyst bed immobilised in the reactor autoclave. This can be retrofitted to a hydrogenation stirred tank reactor which collects the material that has passed through the catalyst bed ahead of its recirculation.

Its limited hydrogen inventory allows more forcing conditions (up to 100 bar and 150°C) to be exploited in flow, with appropriate engineering controls, versus those achievable in batch. Also, as the product is removed from the flow reactor once formed, any overreaction involving the catalyst is likely to be better controlled than in batch. Use of an immobilised bed of catalyst, as in a trickle bed reactor, allows for a high effective molar concentration of catalyst. This is useful where an increase in catalyst concentration drives product formation at the expense of background reactions not involving the catalyst.

Alternatives to the immobilised beds used for continuous hydrogenations are daisychained continuous stirred tank reactors [32], and the Taylor-Couette reactor [33], whose axisymmetric toroidal vortices, providing efficient mixing of a heterogeneous slurry. Neither option allows for the high effective catalyst concentrations of an immobilised bed reactor. As with batch processing, they still require a catalyst filtration operation at the end of the reaction with its impact on cycletime. These are all areas where the use of an immobilised bed can offer strategic advantages, on top of access to processing windows of temperature and pressure that are not available to other technologies, cutting down on the potential exposure of the operator to the catalyst material, and offering a high interfacial area between the liquid and gas phases for hydrogen dissolution. Ideally, catalyst particles of an appropriate size are prepacked into cartridges that can be swapped in and out whenever the catalyst bed needs changing. The penetration of trickle bed technology into the pharmaceutical development and manufacturing services market has been constrained, however, by a lack of catalysts with suitable physical properties. This problem is accentuated further with emerging catalyst-coated static mixer technology [34]. This uses a specialised mixer (doubling as a reagent) with a bespoke manufacture and so its adoption over the use of a commodity like palladium on carbon would require compelling reasons.

7. Conclusions

This chapter has set out principles and workflows that can be adopted when developing a heterogeneous hydrogenation for the development and manufacture of an API or API intermediate. A lean list of experiments has been provided, whose systematic variation of concentration and temperature levers allows the establishing of a working mechanism and catalyst surface chemistry that is kinetically consistent with the observed reaction rate. Even where an initial mechanistic model poorly predicts data gathered in the lab, the designed experimentation used to establish this fact can be recycled for the next mechanistic iteration. Where resource or capability for detailed kinetic studies is constrained, a team can still make useful inferences based on a mechanistic hypothesis, designing experiments that look to establish differences in the driving forces behind activity and selectivity. Similar equity can be extracted by consciously revisiting duty design and equipment requirements at intervals to ensure decisions around manufacturing equipment selection are data-led. Building on the knowledge of the surface chemistry, a reproducible and robust transfer between a donor and receiving vessel, as part of a scaleup or facility transfer, requires consideration around what can prevent reaction solutions in both setups from being in effect saturated with hydrogen throughout their respective courses. It cannot be assumed that the headspace pressure used in the lab will be fit for the pilot or production plant, and that the reaction in the donor facility does not need to be slowed down as part of its accommodation in the receiving facility. Answering these and other points around process accommodation requires the physical characteristics of the manufacturing facility to be considered alongside the kinetics on the catalyst surface, and we have highlighted a number of heuristics to help with this. We have also sought to present a balanced outlook on equipment selection, recognising that their flexibility ensures stirred tank methodology continues to stand the test of time.

Acknowledgements

The author would like to acknowledge discussions with Dr. Felix Roessler and Dr. Neil Hawbaker (Mettler-Toledo), and thanks Mettler-Toledo for access to their DynoChem Resources library.

Author details

Alan Steven
Euroapi, Haverhill, UK

*Address all correspondence to: kjwx109@gmail.com

References

[1] Carey JS, Laffan D, Thomson C, Williams MT. Analysis of the reactions used for the preparation of drug candidate molecules. Organic & Biomolecular Chemistry. 2006;**4**(12): 2337-2347. DOI: 10.1039/b602413k

[2] Roessler F. Catalysis in the industrial production of pharmaceuticals and fine chemicals. Chimia. 1996;**50**(3):106-109

[3] Johnson-Matthey Heterogeneous Catalyst Application Table [Internet]. Available from: https://matthey.com/documents/161599/177860/Hetcat-Kit-Apps-Table.pdf/09a8b6c3-c6a9-b854-5549-1024000c7702?t=1651020978646 [Accessed: July 26, 2024]

[4] Evonik Catalyst Library – Application Table [Internet]. Available from: https://catalysts.evonik.com/en/Services/order-catalysts-samples/attachment/207732?rev=3b2ba59a04cf507a483f53d29229e4e5 [Accessed: July 26, 2024]

[5] Brechtelsbauer C, Hannon J, Hearn S. Characterisation of the Scalability of Hydrogenation Reactions [Internet]. 2002. Available from: https://dcresources.scale-up.com/Open.aspx?file=280&lid=281&c=sus [Accessed: August 07, 2024]

[6] Blaser H-U, Pugin B. The industrial application of heterogeneous enantioselective catalysts. In: Ding K, Uozumi Y, editors. Handbook of Asymmetric Heterogeneous. Catalysis: Wiley; 2008. DOI: 10.1002/9783527623013.ch12

[7] Machado RM. Fundamentals of Mass Transfer and Kinetics for the Hydrogenation of Nitrobenzene to Aniline [Internet]. 2007. Available from: https://www.mt.com/dam/non-indexed/po/autochem/clean/application-notes/hydrogenation-nitrobenzene-to-aniline/fundamentals_of_masstransferandkineticshydrogenation.pdf [Accessed: August 28, 2024]

[8] DynoChem Resources, Mettler-Toledo. Hydrogen Solubility and Heat of Reaction [Internet]. 2019 Available from: https://dcresources.scale-up.com/Open.aspx?file=3288&lid=5564&c=sus [Accessed: July 24, 2024]

[9] Singh UK, Krska SW, Sun Y. Deactivation of heterogeneous hydrogenation catalysts by alcoholic solvents. Organic Process Research and Development. 2006;**10**(6):1153-1156. DOI: 10.1021/op0601520

[10] Tungler A, Szabados E. Overcoming problems at elaboration and scale-up of liquid-phase Pd/C mediated catalytic hydrogenations in pharmaceutical production. Organic Process Research and Development. 2016;**20**(7): 1246-1251. DOI: 10.1021/acs.oprd.6b00073

[11] Crawford CJ, Qiao Y, Liu Y, Huang D, Yan W, Seeberger PH, et al. Defining the qualities of high-quality palladium on carbon catalysts for Hydrogenolysis. Organic Process Research and Development. 2021;**25**(7): 1573-1578. DOI: 10.1021/acs.oprd.0c00536

[12] Masson E, Maciejewski EM, Wheelhouse KMP, Edwards LJ. Fixed bed continuous hydrogenations in trickle flow mode: A pharmaceutical industry perspective. Organic Process Research and Development. 2022;**26**(8): 2190-2223. DOI: 10.1021/acs.oprd.2c00034

[13] Said MB, Baramov T, Herrmann T, Gottfried M, Hassfeld J, Roggan S.

Continuous selective hydrogenation of Refametinib Iodo-nitroaniline key intermediate DIM-NA over Raney cobalt catalyst at kg/day scale with online UV–visible conversion control. Organic Process Research and Development. 2017;**21**(5):705-714. DOI: 10.1021/acs.oprd.7b00039

[14] Carangio A, Edwards LJ, Fernandez-Puertas E, Hayes JF, Kucharski MM, Rutherford GW, et al. Evaluation of sponge metal catalysts in a trickle bed reactor for the continuous hydrogenation of an aliphatic nitro intermediate. Organic Process Research and Development. 2020;**24**(10): 1909-1915. DOI: 10.1021/acs.oprd.9b00447

[15] Argyle M, Bartholomew C. Heterogeneous catalyst deactivation and regeneration: A review. Catalysts. 2015; **5**(1):145-269. DOI: 10.3390/catal5010145

[16] Gleason W, inventor; EI Du Pont de Nemours and Co, assignee. Deactivation of catalysts. United States patent US2810666A. 1957 1957-10-22

[17] LeBlond C, Wang J, Larsen RD, Orella CJ, Forman AL, Landau RN, et al. Reaction calorimetry as an in-situ kinetic tool for characterizing complex reactions. Thermochimica Acta. 1996; **289**(2):189-207. DOI: 10.1016/s0040-6031(96)03072-9

[18] Hinshelwood CN. Kinetics of heterogeneous reactions. In: The Kinetics of Chemical Change. 1st ed. Oxford: Oxford University Press; 1940. pp. 178-233

[19] Horiuti I, Polanyi M. Exchange reactions of hydrogen on metallic catalysts. Transactions of the Faraday Society. 1934;**30**:1164-1172. DOI: 10.1039/tf9343001164

[20] Patil BR, Bari AH, Pinjari DV, Pandit AB. Intrinsic kinetics of three-phase slurry hydrogenation of o-Nitrocardanol to o-Aminocardanol over Raney nickel catalyst. Industrial and Engineering Chemistry Research. 2017; **56**(39):11034-11041. DOI: 10.1021/acs.iecr.7b02523

[21] Laidler KJ. Reactions on surfaces. In: Laidler KJ, editor. Chemical Kinetics. 3rd ed. New York: Harper & Row; 1987. pp. 229-275

[22] Guczi L, Molnár Á, Teschner D. Hydrogenation reactions: Concepts and practice. In: Reedijk J, Poeppelmeier K, editors. Comprehensive Inorganic Chemistry II. Amsterdam: Elsevier; 2013. pp. 421-457. DOI: 10.1016/b978-0-08-097774-4.00713-0

[23] Sun Y, Wang J, LeBlond C, Landau RN, Blackmond DG. Asymmetric hydrogenation of ethyl pyruvate: Diffusion effects on Enantioselectivity. Journal of Catalysis. 1996;**161**(2):759-765. DOI: 10.1006/jcat.1996.0238

[24] DynoChem resources, Mettler-Toledo. Estimate kLa for plant vessel [Internet]. 2024 Available from: https://dcresources.scale-up.com/Open.aspx?file=3309&lid=5585&c=sus [Accessed: August 01, 2024]

[25] Deimling A, Karandikar BM, Shah YT, Carr NL. Solubility and mass transfer of CO and H2 in Fischer—Tropsch liquids and slurries. Chemical Engineering Journal. 1984;**29**(3): 127-140. DOI: 10.1016/0300-9467(84)85038-8

[26] Sun Y, LeBlond C. Understanding fundamental processes in catalytic hydrogenation reactions. In: Am Ende DJ, Am Ende MT, editors. Chemical Engineering in the

Pharmaceutical Industry. 2nd ed. Hoboken: John Wiley & Sons, Inc.; 2019. pp. 191-201. DOI: 10.1002/9781119600800.ch9

[27] Mitchell CW, Strawser JD, Gottlieb A, Millonig MH, Hicks FA, Papageorgiou CD. Development of a Modeling-based strategy for the safe and effective scale-up of highly energetic hydrogenation reactions. Organic Process Research and Development. 2014;**18**(12):1828-1835. DOI: 10.1021/op500207r

[28] Ergun S, Orning AA. Fluid flow through randomly packed columns and fluidized beds. Industrial and Engineering Chemistry. 1949;**4**(6): 1179-1184. DOI: 10.1021/ie50474a011

[29] Kotecki B, Franczyk TS, Allian A, Hanna D, Diwan M. Development of a control strategy for hydrogenation of aryl nitro Analogs by kinetic justification. Organic Process Research and Development. 2023;**27**(9): 1616-1623. DOI: 10.1021/acs.oprd.3c00085

[30] Stitt EH, Fishwick RP, Natividad R, Winterbottom JM. Multiphase hydrogenation reactors-past, present and future. In: Jackson SD, Hargreaves J, Lennon D, editors. Catalysis in Application. London: The Royal Society of Chemistry; 2003. pp. 153-160. DOI: 10.1039/9781847550347-00153

[31] Roessler F. Catalytic hydrogenation in the liquid phase. Chimia. 2003;**57**(12): 791-798. DOI: 10.2533/000942903777678416

[32] Alsten JGV, Jorgensen ML, am Ende DJ. Hydrogenation of a pharmaceutical intermediate by a continuous stirred tank reactor system. Organic Process Research and Development. 2009;**13**(3):629-633. DOI: 10.1021/op800170r

[33] Schrimpf M, Esteban J, Warmeling H, Färber T, Behr A, Vorholt AJ. Taylor-Couette reactor: Principles, design, and applications. AIChE Journal. 2021;**67**:e17228. DOI: 10.1002/aic.17228

[34] Legg R, Zhang C, Bourchier M, Cole S, Martinez-Botella I, Nguyen X, et al. Durability study of 3D-printed catalytic static mixers for hydrogenations in chemical manufacturing. Chemie Ingenieur Technik. 2022;**94**(7):1017-1023. DOI: 10.1002/cite.202200060

Chapter 8

Utilization of Red Mud Waste as a Dye Adsorbent

Hellna Tehubijuluw, Didik Prasetyoko, Yuly Kusumawati, Riki Subagyo and Reva Edra Nugraha

Abstract

Bauxite waste (Red Mud) from Bintan Island Indonesia, Riau Archipelago, is used to synthesize hierarchical ZSM-5. This book chapter discusses the preparation of a hierarchical ZSM-5 material as an adsorbent for methylene blue removal from water using the hydrothermal method. The synthesis process involved two stages of crystallization: The first stage was conducted at 80°C for 6, 9, and 24 h, followed by the second stage at 150°C for 24 h. The findings revealed that hierarchical ZSM-5_6h possessed a high mesoporous surface area and volume, measuring 734 m^2/g and 0.58 cm^3/g, respectively. The adsorption kinetics of hierarchical ZSM-5_6h on methylene blue adhered to a pseudo-first-order model. The thermodynamic parameters, including ΔG°, ΔH°, and ΔS°, were − 2.69 $kJmol^{-1}$; −4,04 $kJmol^{-1}$; −5.38 $kJmol^{-1}$, 38.06 $kJmol^{-1}$, and 0.13 kJ/mol.K, respectively.

Keywords: adsorption, bauxite waste, methylene blue, hydrothermal method, hierarchical ZSM-5

1. Introduction

1.1 Overview of Red Mud waste and its conversion into adsorbent materials

In today's world development, industrialization is an important process that occurs all over the world. This process is related to the large environmental impact related to safe and appropriate waste disposal and waste management generated. One of the wastes produced is Red Mud (RM) [1, 2]. RM is waste in the form of sludge which is a by-product of bauxite ore processing. The bauxite ore processing process produces 1.0–1.8 tons of RM per ton of aluminum produced [3]. With the increasing world demand for aluminum, the amount of RM produced as a by-product of aluminum processing is increasing. The overall RM inventory is estimated to reach 4 billion tons in 2019, with a production rate of 0.15 billion tons per year [4].

RM has high alkalinity, in dry form with a pH ranging from 10.5 to 12.5 [5]. Large amounts of RM disposal cause environmental problems such as soil contamination, groundwater pollution, and the formation of suspensions in the waters. The accumulation of dry RM around bauxite processing plants also causes dust pollution, which is also a serious health problem for the ecosystem [6]. To overcome the environmental problems caused by the accumulation of RM, especially in wastewater, is to utilize RM as a basic material for the synthesis of other materials that have utility.

Several studies have been conducted by utilizing RM as a catalyst [7, 8], and the basic material for making building materials [3, 9], inorganic substrates [10], the development of dyes and paints [11], and as adsorbents, especially to overcome water pollution problems [12–16].

Optimizing the use of RM as an adsorbent is carried out by converting it into other materials that have a higher surface area such as mesoporous materials. Several researchers have converted RM waste as a source of alumina and silica in the synthesis of other materials such as mesoporous aluminate [17], mesoporous alumina [18], and ZSM-5 hierarchical [19]. The most oxide compounds in RM are Fe_2O_3, Al_2O_3, and SiO_2 [18]. The high content of Al_2O_3 and SiO_2 has the potential to be used as a precursor to ZSM-5, while hematite (Fe_2O_3) must be separated first so as not to affect the purity of the synthesis of aluminosilicate. Zinoveev et al. [20] developed a direct combustion technology with sodium carbonate reduction then with magnetic separation of high iron content in RM. The results obtained through this process are 90.87% iron, and the total iron as a whole is 95.76% [20]. Microwave reduction method is used to extract iron from iron minerals in RM [21, 22]. The study showed that the reduction of RM with *microwave* at 1000°C for 10 min obtained an iron concentration of 48.5% and an iron purity level of 95%. The separation of Fe_2O_3 from RM has also been carried out by Bahari et al. [18]. The separation is carried out based on physical properties where it is known that the specific gravity of Fe_2O_3 is greater than that of Al_2O_3 and SiO_2. However, this separation method is ineffective because Fe_2O_3 has not been completely separated, so the alumina extraction process must be continued with a hydrometallurgical method that requires chemical reagents (HCl and NaOH) in large quantities. Another method that can be used to separate Fe_2O_3 into the extraction of alumina and silica from RM is by the alkali fusion method. The principle of alkali fusion reaction is the reaction between alumina and silica with NaOH alkaline base at high temperature (about 550–650°C). During the heating process, alumina and silica will fuse with NaOH to form sodium aluminate and sodium silicate salts that are easily soluble in water, making them easier to separate [23].

The extraction of alumina and silica by the alkali fusion method has been carried out by Lin et al. [24] to extract alumina and silica from coal fly ash of power plants and used to synthesize sodalite and silica. The optimum condition for silica extraction from fly ash was 46.2% at a temperature of 95 °C for 3 h. Sodalite is synthesized at 200 °C for 1 hour. The silica produced has a purity of 98.8% [24]. Zhang et al. [25] successfully synthesized magnetic zeolite from a mixture of RM and fly ash using the alkali fusion method with NaOH (NaOH ratio: RM-fly ash 1.2) at a temperature of 600°C to extract alumina and silica from the mixture which was then continued by hydrothermal method [25]. Padhi et al. [17] reported that Fe_2O_3 in RM Bintan Island can be separated from Al_2O_3 and SiO_2 through alkali fusion reaction. The addition of NaOH to the fusion alkali is carried out with a ratio of NaOH mass: RM 1:1 and heating at a temperature of 600°C. Chen et al. [7] performed a heating temperature variation to determine the temperature with the highest amount of Al_2O_3 and SiO_2 obtained after the alkaline fusion of RM Bintan Island. The reported optimum temperature is 450°C [7]. Zhao et al. [19] also performed alkali fusion with a method that has been optimized by Padhi et al. [17] and Chen et al. [7]. RM Bintan Island plus NaOH (NaOH mass ratio: RM 1:1) and heated at 450°C for 2 h [19].

Temperature and crystallization time are important factors that can affect the process of zeolite synthesis. Sun et al. [26] have synthesized ZSM-5 using hydrothermal methods and crystallization times ranging from 24 to 96 h at a temperature of 180 °C [26]. Aloui et al. [27] have successfully synthesized ZSM-5 from kaolin without organic templates directly. The variations in the crystallization time used were 12, 24, 48, and 72 h. About 24 h is the optimal time to obtain ZSM-5 crystals with high crystallinity in the form of hexagons measuring 0.57–2.57 μm [27]. Aguirre-Cruz et al. [28] succeeded in synthesizing ZSM-5 hierarchical from kaolin as a source of silica and alumina, and silicate seeds were added as a structural director of ZSM-5 and CTAB as a structural director of meso-sized pores. The ZSM-5 hierarchy was synthesized at a crystallization temperature of 80; 127.5; and 175°C and the crystallization temperature varied in crystallization time of 12, 24, and 48 h, respectively. The results showed that samples with a crystallization time of 48 h at a temperature of 80°C had a high surface area and mesopores [28].

In this study, the synthesis of zeolite ZSM-5 (Zeolite Socony Mobil-5) hierarchical from RM has been carried out on Bintan Island, Riau Islands. The first stage of the crystallization process uses a hydrothermal method at a temperature of 80°C with a time variation of 6, 9, and 24 h. At this stage, mineralizing agents play an important role in the formation of Si-O-Al bonds and produce clusters (ZSM-5 precursors) that are nanocrystalline in size [29]. These nanoclusters will then undergo arrangement into larger aggregates. The second stage is 150 °C for 24 h. The second crystallization process is the crystallization stage of the formation of meso-sized pores where there is an interaction between surfactants and aluminosilicate nanoparticles. The selection of hierarchical ZSM-5 synthesis in this study is because the pore diameter of ZSM-5 is in the range of 0.54–0.56 nm, where the pore size is too small and blocks the diffusion process of large reactants on the active side of the catalyst [30]. Moreover, this approach is taken because the uniform distribution of zeolite micropores influences the selectivity in various reactions. However, it can also lead to diffusion constraints when large molecules navigate through relatively narrow channels. Some active sites are prevented from connecting reactants due to the limited diffusion process. As a result, it is indispensable to develop new materials with meso-sized porosity. Several solutions have been proposed to overcome the limitations of diffusion, one of which is the synthesis of new zeolites with larger pores, such as the hierarchical ZSM-5 synthesis, which modifies the pore structure in zeolite by adding a mesoporous structure direction template [31] and also modifies the micro-pore ZSM-5 to ZSM-5 with hierarchical pores that have combined pores between micro and meso so that reactants with large sizes can enter the pores and be diffused in the active side of catalysts [32, 33].

The technology of using adsorbents in overcoming water pollution problems has many limitations, including the need for additional processes to separate adsorbates from adsorbents and further handling of adsorbed pollutants [34–36]. This research is the utilization of RM Bintan Island, Riau Islands, which is left to be wasted into the environment. The waste is processed into zeolite (ZSM-5 hierarchy) by hydrothermal method of two stages of crystallization at a temperature of 80°C at a time of 6, 9, and 24 h and a temperature of 150°C at a time of 24 h. Furthermore, an adsorption study was carried out to test the activity of ZSM-5 hierarchical as an adsorbent so that it could remove MB in water.

2. Synthesis and characterization of ZSM-5 hierarchy as blue methylene adsorbent in water

This section contains a description of the RM preparation from Bintan Island, Riau Islands, which is used for the synthesis of hierarchical ZSM-5 with the addition of

CTAB surfactant as a meso-pore directing template, the organic compound tetrapropyllammonium hydroxide (TPAOH) as the MFI structure director, and LUDOX as a silica additive by hydrothermal method as well as the adsorption study of hierarchical ZSM-5 as an MB adsorbent in water.

2.1 RM preparation and characterization

The RM used in this study came from Riau Island, Bintan Island, in the form of dry chunks. The chunks are mashed with a mortar and then filtered with a 325-mesh sieve. The sieving aims to ensure that the RM sample particles used have a uniform size and to accelerate the reaction between RM and NaOH at the time of fusion reaction. Next, the RM is dried at a temperature of 105°C, which aims to remove the water that is physically bound to the RM. RM has a different compound content depending on the region of origin of bauxite and its processing technique [37] so it is necessary to carry out initial characterization to determine the phase and chemical composition of RM. In this study, RM from Bintan Island was characterized by XRF, XRD, and FTIR instrumentation. The composition of oxide compounds in RM is known from the results of characterization with XRF instruments.

Based on the results of XRF analysis, the dominant compound content in RM used in this study is iron oxide compounds (Fe_2O_3), alumina (Al_2O_3), and silica (SiO_2), and the rest are minor compounds, namely TiO_2, P_2O_5, K_2O, CaO, CuO, V_2O_5, Cr_2O_3, ZrO_2, MnO, ZnO, and PbO. The levels of compounds contained in RM Bintan Island are shown in **Table 1**.

The mineral content of RM was also analyzed using XRD techniques. XRD patterns are used to identify the minerals present in RM. RM generally contains minerals that have not been dissolved in the Bayer process such as hematite (Fe_2O_3), goethite, (FeOOH), gibsite ($Al(OH)_3$), quartz (α-SiO_2), anatase (β-TiO2), and baumit (γ-AlO (OH) [3].

Based on JCPDS and ICSD data, RM consisting of gibsite mixture (JCPDS 00–007-0324) has a diffraction peak at 2θ = (18.28; 20,31; 20,54; 36,59; 37.64°), buhmit

Oxide compounds	Weight (%)
Fe_2O_3	39.22
Al_2O_3	30.00
SiO_2	25.70
TiO_2	2.96
CaO	0.30
ZrO_2	0.26
PbO	0.22
K_2O	0.14
CuO	0.13
V_2O_5	0.08
ZnO	0.04
Cr_2O_3	0.03

Table 1.
RM composition from Bintan Island.

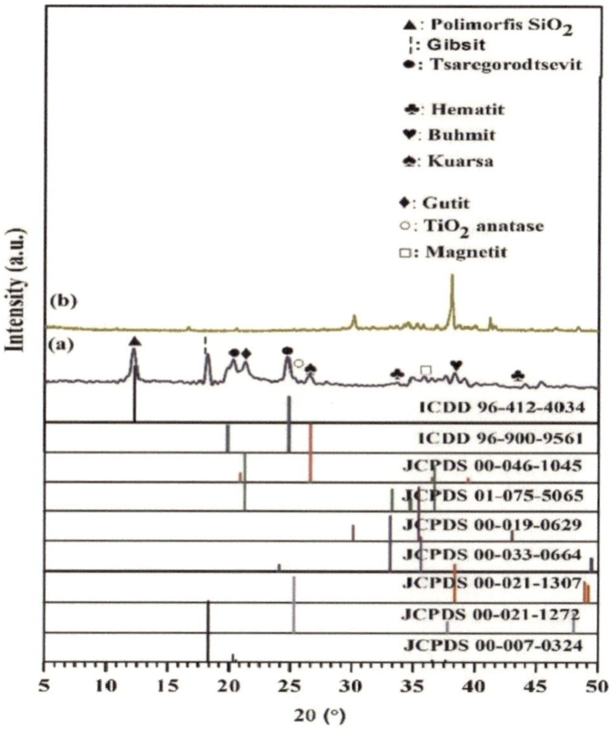

Figure 1.
X-ray diffraction pattern from (a) RM waste, (b) alkaline fusion filtrate (FAF) and gibsite JCPDS data ⬟(00-007-0324), TiO_2 anatase ○(00-021-1272), buhmit ♥ (00-021-1307), hematite ♣ (00-033-664), magnetite □ (00-019-0629), goethite ◆ (01-075-5065), kuarsa ♠ (00-046-1045) dan data ICSD tsaregorodtsevite • (96-900-9561) dan polimorfis silika ▲ (96-412-4034).

(JCPDS 00–021-1307), 2θ = (38.38; 49.93; 49.21°), hematite (JCPDS 00–033-0664), 2θ = (24.14; 33.15; 35.61; 40,86; 49,48; 54.90°), magnetite (JCPDS 00–019-0629), 2θ = (30.91; 35.42; 43.04), goethite (JCPDS 01–075-5065), 2θ = (21.27; 33,28; 34,74; 36.72°), quartz (JCPDS 00–046-1045), 2θ = (20.86; 26,64; 36,54; 39.47°), TiO_2 anatase (JCPDS 00–021-1272), 2θ = (25.28; 37.80; 48.05°), ICSD data for tsaregorodtsevite (96–900-9561), 2θ = (19.84; 24.88°), and SiO_2 polymorphism (96–412-4034) 2θ = (12.29°). The XRD pattern of RM and the minerals present in RM are shown in **Figure 1**.

Characterization by FTIR spectroscopy was carried out to determine the functional groups and types of bonds present in the RM sample. The infrared spectrum of RM is shown in **Figure 2**. The FTIR RM spectrum shows the absorption bands of quartz and hematite at wave numbers 999–1008, 549–540, and 462–468 cm^{-1} ([38]. The absorption band at the wave number of 960 cm^{-1} indicates the presence of Si-O bonds from quartz. Furthermore, the presence of hematite absorption bands at wave numbers 559 and 479 cm^{-1} shows Fe^{3+}-O_2^- bonds. The absorption bands that appear at wave numbers 874, 712, and 628 cm^{-1} indicate the presence of Al^{3+}-O_2^- bonds [39].

2.2 Alkaline fusion reaction of RM with NaOH

RM is a source of Al and Si required in the synthesis of zeolite materials because the content of Al_2O_3 and SiO_2 is quite high. The main composition of RM besides Al_2O_3 and SiO_2 is Fe_2O_3. The Fe_2O_3 content in this study was even higher than the two

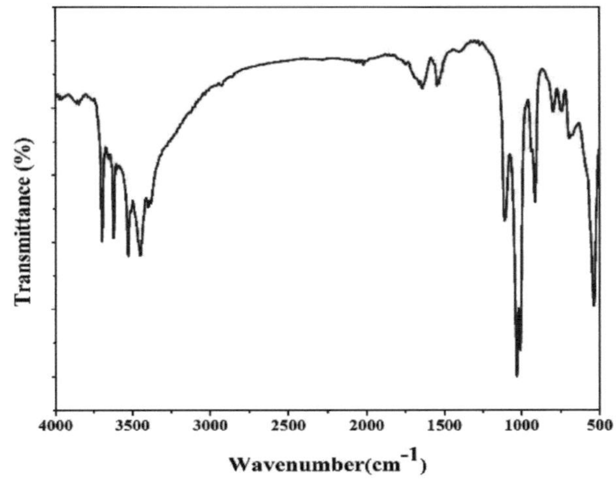

Figure 2.
Spectra FTIR RM Bintan Island.

Oxide compounds	Heavy (%)	
	AFF	AFR
Fe_2O_3	0.50	45.11
Al_2O_3	61.50	21.00
SiO_2	20.50	28.20
TiO_2	—	3.147

Table 2.
XRF data comparing the composition of the main compounds FAF and RAF reactions.

compounds mentioned earlier. Bahari et al. [18] reported that Fe_2O_3 in RM Bintan Island can be separated from Al_2O_3 and SiO_2 through an alkaline fusion reaction. Alkaline fusion is a method used to separate Fe_2O_3 in the extraction of alumina and silica from RM [40]. In addition, it aims to decompose and activate the SiO_2 and Al_2O_3 components in RM into a form of salt that has higher solubility in water.

The solids obtained from the alkali fusion filtrate process are characterized by XRF. **Table 2** shows the composition of the main compounds in the reaction of alkali fusion filtrate (AFF) and alkali fusion residue (AFR). The high Fe_2O_3 content in AFR (45.11%) resulted in a brownish-red residual color compared to the RM color before alkali fusion (39.22%), as reported by Gök et al. [41]. TiO_2 also only appears in the results of XRF characterization of residues. This shows that the residue of the alkaline fusion process has metal oxides that are difficult to dissolve in water under alkaline conditions [42].

In this study, sodium hydroxide (NaOH) was used as an alkaline source. NaOH functions as a mineralizer that can increase solubility by forming sodium aluminate salts and sodium silicates [25].

Furthermore, the alkaline fusion method that has been optimized by Padhi et al. [17] and Ramdhani et al. [43] is used for the alkaline fusion reaction of RM Bintan

Island by adding NaOH (NaOH mass ratio: RM 1:1) and then heated at a temperature of 450°C for 2h. The reaction that occurs during the alkaline fusion reaction between SiO_2 and Al_2O_3 components with NaOH is as follows [44]:

$$2NaOH_{(s)} + Al_2O_{3(s)} \rightarrow 2NaAlO_{2(s)} + H_2O_{(g)} \quad (1)$$

$$2NaOH_{(s)} + SiO_{2\ (s)} \rightarrow 2Na_2SiO_{3(s)} + H_2O_{(g)} \quad (2)$$

Eqs. (1) and (2) describe the process of alkaline fusion of Al_2O_3, and SiO_2 reacts with NaOH to form $NaAlO_2$ and Na_2SiO_3 salts. This is shown in the FTIR spectrum of the emergence of new absorption peaks at wave numbers of 870 and 1420 cm^{-1} indicating the presence of sodium aluminate [45] and sodium silicate [46, 47]. Furthermore, in the alkaline fusion process, there is no Fe-O bond, indicating that the iron oxide in RM has been lost (**Figure 3**).

Kusumawardani et al. [48] reported that the salt produced from the alkali fusion process can be dissolved in water, while iron oxide does not so that AFF will be produced with the content of alumina and silica salts. This is in accordance with the XRF results from the FAF where the Fe_2O_3 content is small (0.50%) but has a large Al_2O_3 content (61.50%). The SiO_2 content in AFR is quite large (28.20%) compared to the content in AFF (20.50%). This is suspected to be due to the concentration of NaOH that is too low to dissolve all the SiO_2 and Al_2O_3 content in RM. Liu et al. [49] reported that the solubility of SiO_2 and Al_2O_3 increased along with the increase in the concentration of NaOH used and the solubility of Al_2O_3 was greater than that of SiO_2 in alkaline conditions [49]. XRF results show that the Al_2O_3 content in AFF is greater than the SiO_2 content [48].

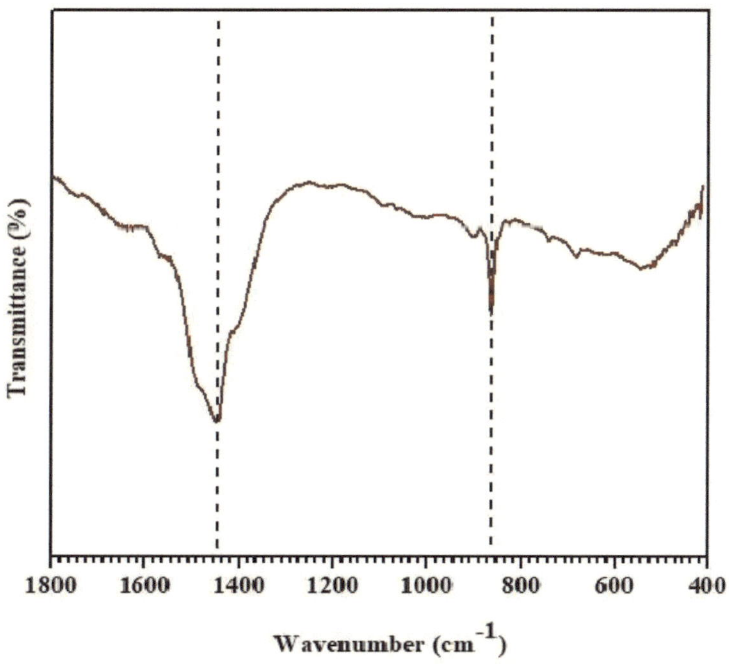

Figure 3.
FTIR RM spectra after alkali fusion.

2.3 Sintesis ZSM-5 hierarki dari RM

The hierarchical synthesis of ZSM-5 from AFF results from the alkaline fusion process of RM, namely sodium aluminate-sodium silicate. Based on the results of XRF, AFF sodium aluminate-sodium silicate obtained Al and Si levels of 61.5 and 20.5%, respectively. The Al and Si content data were used to determine the molar ratio of the ZSM-5 synthesis hierarchy. The hierarchical ZSM-5 synthesis in this study refers to the synthesis method from the research of Nguyen et al. [50] and Zhao et al. [19], which was modified using the hydrothermal method and then added CTAB from the Gao et al.'s [51] research. The composition of the molar ratio used by Gao et al. [51] and Zhao et al. [19] are $100SiO_2$: $2Al_2O_3$: $1800H_2O$: 20TPA. In this study, 1/500 of the original ratio was used so that the composition of the molar ratio used was 0.2 SiO_2: 0.004 Al_2O_3: 3.6 H_2O: 0.04 TPA.

The Al content in AFF is more than the Si content. Therefore, Si is added from other sources to meet the needs of Si to meet the desired composition of synthetic molar (0.2 SiO_2: 0.004 Al_2O_3). This study uses colloidal silica (LUDOX) as a source of silica. Other materials used are TPAOH as the MFI structure directing agent, CTAB as the mesoporous structure directing agent, and DM aqua as the H_2O source.

The process of synthesis of hierarchical ZSM-5 from RM Bintan Island Indonesia began by weighing the AFF resulting from alkali fusion of 0.663 g and dissolved with DM water of 24.668 mL in a polypropylene bottle. The hierarchical ZSM-5 synthesis process requires a large amount of SiO_2, while the SiO_2 content in the FAF from alkali fusion is only 20.50%, so it is necessary to add SiO_2 from other sources. LUDOX is one of the sources of SiO_2 that can be used in the synthesis process of ZSM-5 [51]. The addition of LUDOX of 39.901 g was carried out after the FAF results from complete dissolved alkali fusion and was done slowly and interspersed with the addition of DM aqua so that clumps would not form. The mixture is stirred with a magnetic stirrer for 8 h at room temperature. This is done so that the particles are evenly distributed. The role of DM fish is as a solvent to improve mixing and material transfer [52], change the physical/chemical properties of reactants and products, and accelerate reactions [53]. Furthermore, 20.337 g of TPAOH was added to the pp. bottle (gel mixture) and stirred for 30 min at room temperature.

The hydrothermal process is carried out at a temperature of 80°C for 6 h. This process occurs continuously, which begins with a condensation reaction and is followed by polymerization of saturated solution to form Si-O-Al bonds. During the hydrothermal process, an amorphous Si-O-Al bonding network containing water occurs and produces nanocrystalline aluminosilicate clusters [54]. After the first stage of the hydrothermal process is completed, the crystallization process is stopped by cooling the reaction vessel (polypropylene bottle) using running water. This sudden cooling is carried out to prevent the formation of perfect aluminosilicate crystals with a larger size and smaller pore size [51]. This circumstance is avoided because it can make it difficult to combine these particles in the mesoporous formation process, which is assisted by the CTAB template in the next process. Mesoporous formation is carried out by adding surfactants (CTABs) to the resulting white gel. CTAB (Molar ratio SiO_2/CTAB = 3.85) is added slowly, and the distiller is strong for 1 h. The stirring gel mixture is transferred into the autoclave for a second stage of crystallization at 150 °C for 24 h through structural arrangement and micelle formation until mesopores are formed in the solid. The same synthesis process is carried out for hydrothermal processes at 80 °C for 9 and 24 h. The resulting solids are then washed with DM aqua to pH neutral and dried at 60°C for 24 h to remove the moisture content. Drying is carried out at a low temperature so that the water comes out slowly so as not to damage the mesoporous structure formed because the Si-O-Al skeleton is still very fragile [55]. The solids formed are then calcined at a temperature of 550 °C for 1 hour with a temperature increase of 2 °C/min using a stream of nitrogen gas so that the decomposition of the template takes place slowly so as not to damage the skeletal structure and strengthen the formation of brittle aluminosilicate (Si-O-Al) bonds [56]. It is then calcined at 550°C for 6 h in free air to remove organic templates (carbon) and strengthen the Si-O-Al network. Melek et al. [57] stated that the CTAB decomposition process takes place in three stages, namely the release of ammonia molecules at 200–320°C which is the result of partial decomposition of CTAB, the breaking of the CTAB template structure at 320°C, and the removal of CTAB from the solid skeleton at 395°C [57]. The calcinated solids were then characterized by XRD, FTIR, SEM-EDX, and nitrogen adsorption-desorption.

2.4 Hierarchical ZSM-5 characterization

2.4.1 Characterization by X-ray diffraction (XRD)

Characterization with XRD technique is based on a sample that has a crystalline field. XRD characterization techniques are used to determine the crystalline phase, crystal structure, and particle size of materials such as ceramics, polymers, composites, and other materials [58]. In this study, XRD analysis was carried out at an angle of 2θ between 5° and 50°. The peak diffraction characteristics of the commercial ZSM-5 and the hierarchical ZSM-5 synthesized at 6, 9, and 24 h are shown in **Figure 4**.

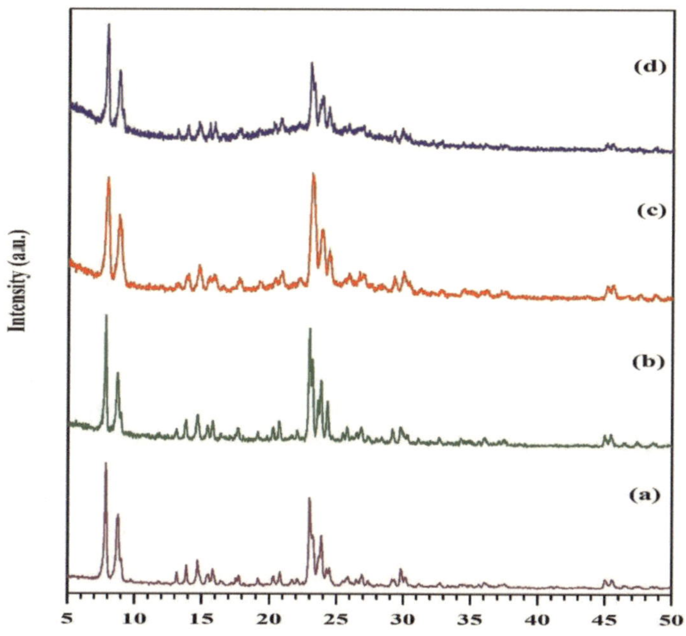

Figure 4.
XRD pattern of samples (a) commercial ZSM-5, and ZSM-5. Synthesis results: (b) ZSM-5_6 h, (c) ZSM-5_9 h, and (d) ZSM-5_24 h.

Intensity Top (°)	**Sample**			**Commercial ZSM-5**
	6h	**9h**	**24h**	
7.9	774.41	773.83	1091.27	2374.61
8.8	510.87	563.97	560.93	1447.73
22.9	575.49	707	1071,12	1912.99
23.15	574	825	918	1009.32
23.29	353.61	559.12	603.21	1135.61
24.3	265.75	419	454	408.18
Total	2480.13	2721.92	4698.53	8488.44
Crystallinity Relative (%)*	29.21	32.07	55.35	100.00

**Relative crystallinity is calculated from the sum of the highest source intensities of the sample compared to the sum of the highest standard peak intensities.*

Table 3.
Relative crystallinity of ZSM-5 hierarchy and commercial ZSM-5 standard.

The distinctive diffraction peaks of ZSM-5 with the MFI structure appeared at 7.9, 8.8, 22.9, 23.1, 23.9, and 24.3°, corresponding to the hydrothermal synthesis durations of 6, 9, and 24 h [59]. Relative crystallinity is calculated from the sum of the intensity of each sample divided by the sum of the intensity of the standard [60]. The calculation of relative crystallinity is based on the six characteristic peaks of the ZSM-5. The commercial ZSM-5 was chosen as the standard because it had the highest intensity compared to other samples. Based on **Table 3**, it shows that the longer the crystallization time and the higher the crystallization temperature, the crystallinity of the sample will increase. The highest crystalline was shown in the ZSM-5_24 h sample, while the lowest was the ZSM-5_6h sample, which was 55.35 and 29.21%, respectively. During crystallization, the formation of zeolite from the solution phase to the solid phase forms a Si-O-Al bond. The crystal growth process of ZSM-5 also occurs during the crystallization process. The crystallinity will certainly continue to increase along with the increase in the crystallization time because the ZSM-5 crystals that are formed will also increase [61]. Therefore, the longer the crystallization time produces the intensity and percentage of crystallinity of the ZSM-5 sample, the greater the percentage.

In **Table 4**, it shows the highest crystallinity of ZSM-5 formed at 24 h when compared to the crystallinity at 6 and 9 h. This difference in intensity is expected to occur. Variations in the number of crystal fields formed in each sample result in different amounts of light reflected from the crystal planes. Samples with higher reflectivity will produce greater intensity, leading to increased crystallinity in the synthesized ZSM-5. Based on the results of XRD, the intensity value increases with increasing crystallization time, indicating high crystal formation of ZSM-5. Furthermore, the ZSM-5 properties of the synthesized hierarchy of RM are shown in **Table 4**.

2.4.2 Characterization by Fourier transform infrared (FTIR) spectroscopy

Fourier transform infrared (FTIR) spectroscopy is a characterization technique used to determine the functional groups present in a compound by utilizing the vibration of the functional groups that are subjected to infrared radiation. According

DOI: http://dx.doi.org/10.5772/intechopen.1010397

Sample	Crystallinity (%)	Particle size (μm)[1]	SBET (m^2/g)	Smeso (m^2/g)		Smicro (m^2/g)[4]	Vtotal (cm^3/g)	Vmicro (cm^3/g)	Vmeso (cm^3/g)		dmicro (nm)	dmeso (nm)	
				Inter-[2]	Intra-[3]				Intra-[2]	Inter[3]		Intra-	Inter-
ZSM-5_6h	29.21	2.12	734	252	562	233	0.85	0.25	0.58	0.023	1.18	3.02	27.5;49.2
ZSM-5_9h	32.07	3.52	721	223	590	261	0.75	0.20	0.53	0.021	1.18	.2.79	298;47.5
ZSM-5_24h	55.35	2.68	388	112	355	238	0.39	0.16	0.21	0.018	1.24	3.32	27.5;49.3
ZSM-5_6 jam Desorpsi	6	—	683	276	564	204	0.83	0.25	0.57	—	1.47	2.89	—

[1]*determined by SEM analysis.*
[2]*S_{meso}, V_{meso} determined using the NLDFT method within a relative pressure range of 0.9–1.0.*
[3]*S_{meso}, V_{meso} estimated using the NLDFT method within a relative pressure range of 0.24 to 0.9.*
[4]*S_{micro} is evaluated using the t-plot method.*

Table 4.
Properties of ZSM-5 hierarchical synthesized from RM.

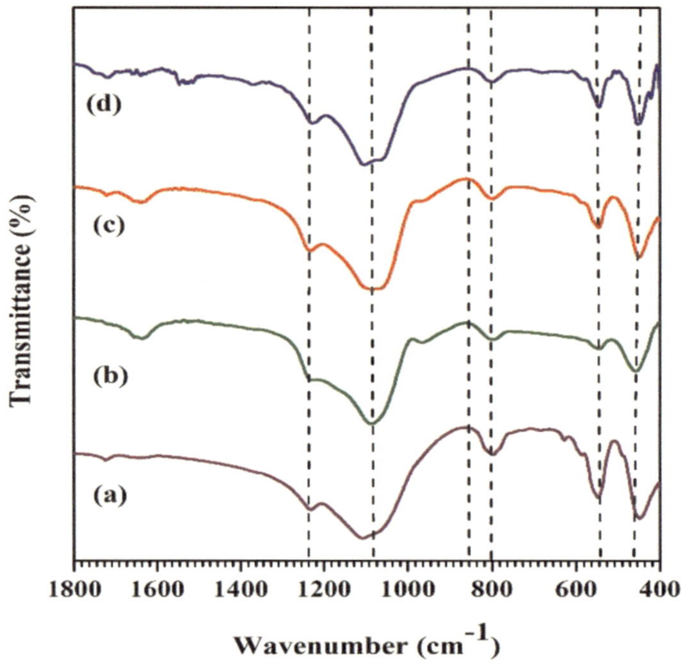

Figure 5.
FTIR spectra (a) commercial ZSM-5, (b) ZSM-5_6 h, (c) ZSM-5_9 h, and (d) ZSM-5_24 h.

to Zhou et al. [62], FTIR can provide information in estimating the molecular structure and functional groups of compounds in the characterized solids [62]. In this study, characterization with FTIR technique was used to determine the peak of characterization of ZSM-5.

The results of the hierarchical ZSM-5 synthesis in **Figure 5** show that the presence of T-O-T bending vibrations (T is an Al or Si atom) appears at a wave number of 450 cm^{-1}, while the absorption bands of 798 and 1100 cm^{-1} are vibrations related to sensitive external tetrahedral structures (external symmetrical tendrial vibrations) and sensitive internal tetrahedral structures of the TO_4 asymmetric longitudinal vibration model.

These absorption peaks are insensitive to changes in tetrahedral structure in ZSM-5 [63]. The absorption band of 540 cm^{-1} arises due to the vibration of the MFI's typical pentacil double ring, which shows the characteristics of MFI-type zeolite pentacils. The absorption band of 1220 cm^{-1} shows an asymmetrical tensive vibration model of T-O-T (tetrahedral, TO_4) that is sensitive to structural changes from tetrahedral external vibrations. Based on **Figure 5**, it can be concluded that the intensity increases with the increase of crystallization time, resulting in the formation of a high crystal structure of ZSM-5 [64].

2.4.3 N_2 adsorption

The nitrogen adsorption-desorption technique is a technique used to determine the distribution of pore size and specific surface area of a solid. The techniques used for hierarchical materials are BET and BJH. The BET (S_{BET}) method is used to determine the specific surface area, while the BJH method is used to determine the pore size distribution.

The adsorption-desorption isotherm N_2 in **Figure 6** shows the isotherm of the sample ZSM-5 at the variation in crystallization time. Rapid adsorption of N_2 at

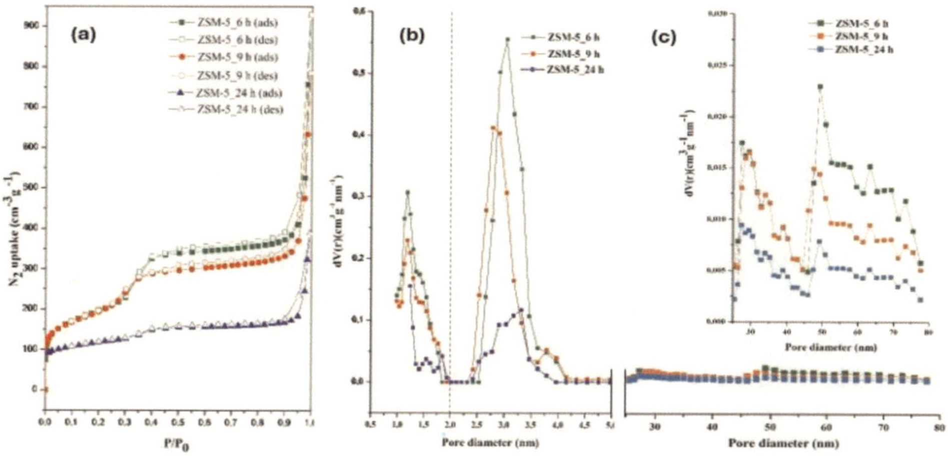

Figure 6.
N_2 adsorption-desorption isotherm (a) and NLDFT pore size distribution of all synthesized samples. The dashed line at 2 nm indicates the boundary of micropores and mesopores (b), the pore size distribution of NLDFT for interparticle mesopores (P/P_o = 0.9–1.0) (c).

relatively low pressure (P/P0) is characteristic of micropore adsorption (type I) and occurs in all samples. An increase in adsorption at a relatively high pressure (P/P0) indicates the presence of mesoporous adsorption (Type IV).

The ZSM-5_6h sample exhibited N_2 adsorption at a high relative pressure (P/P_0) of 0.3–0.9, followed sequentially by ZSM-5_9h and ZSM-5_24h. The findings indicated that shorter crystallization times resulted in ZSM-5 with higher mesoporosity. The appearance of a hysteresis loop, indicating nitrogen molecule adsorption in small quantities, was observed at a relative pressure (P/P_0) starting from 0.3 in both ZSM-5_6 h and ZSM-5_9 h samples. This indicates the occurrence of condensation in the intra-particle mesopores [65, 66]. Meanwhile, *the* hysteresis loop that occurs in ZSM-5_24 h is a high number at a relative pressure *of* P/P0 = 0.9–1.0 indicating the formation of inter-particle mesopores in the structure of ZSM-5 [65, 66].

The volume of mesopores and surface area decreased with the length of crystallization time so that the formation of ZSM-5 hierarchical in the zeolite structure decreased [67]. **Figure 6b** shows the distribution of ZSM-5 pore size obtained from the crystallization time data, calculated using the non-local density functional theory (NLDFT) method, and then shown in **Table 4**. All ZSM-5 samples showed a micropore distribution of ~1.18–1.24 nm; however, the pore volume increased for the ZSM-5_6 h synthesized in a shorter time. The presence of intra-particle mesopores is shown to be between 2.5 and 4.0 nm. High N_2 adsorption volumes are present at ZSM-5_6 h, as well as at ZSM-5_9 h and ZSM-5_24 h. In **Figure 6c**, the pore size distribution in the inter-particle mesopores is calculated at P/P_0 = 0.9–1.0 using the NLDFT method. Furthermore, the volume uptake of N_2 is reduced in large pores, so that the inter-particle pores in the ZSM-5 structure are lower than those of intra-particle pores. These results prove the effect of crystallization time on the formation of hierarchical ZSM-5 from RM waste.

2.4.4 Morphological analysis using SEM and TEM

SEM characterization is used to determine the surface morphology of solid samples. The morphology and particle size of the sample were observed by

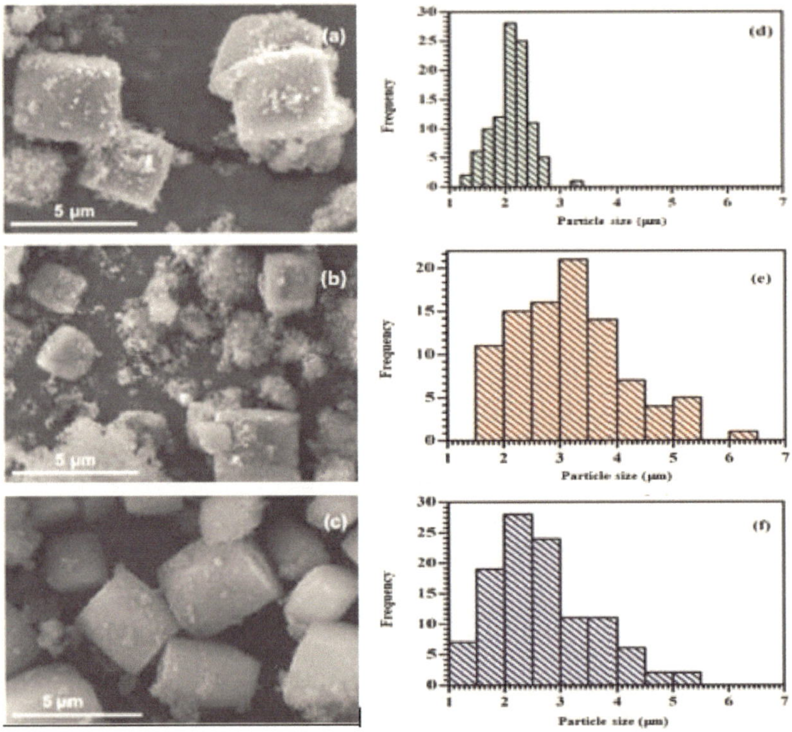

Figure 7.
SEM micrograph of (a) ZSM-5_6 h, (b) ZSM-5_9 h, (c) ZSM- 5_24 h. Size distribution of ZSM-5 particles produced at hydrothermal time (d) 6 h, (e) 9 h, and (f) 24 h.

scanning electron microscopy (SEM). The samples of ZSM-5_6 h, ZSM-5_9 h, and ZSM-5_24 h were tested with SEM to determine the effect of crystallization time on the morphology of solids, as mentioned in Section 2.4.1 each sample had a difference in crystallinity based on the XRD results of 29.21; 32.07; and 55.35%.

In general, the morphology of the three ZSM-5 samples of the microcrystalline hierarchy is cubic (**Figure 7a-c**). Irregular aggregates are dispersed on the surface of the cube microcrystalline at ZSM-5_6 h and ZSM-5_9 h. The increase in crystalline occurs because of an increase in the number of irregular particles present in the sample. These irregular particles are amorphous aluminosilicate particles that arise as a result of the reaction between alumina and silica [68]. This is in accordance with the XRD analysis data in **Table 4** showing that a longer crystallization time will increase the crystallinity of the ZSM-5 hierarchy. The histogram of the particle size distribution is shown in **Figure 7d-f**.

The particle sizes of ZSM-5_6 h, ZSM-5_9 h, and ZSM-5_24 h were 2.12, 3.52, and 2.68 μm, respectively. HRTEM analysis was employed to examine the morphology of ZSM-5_6 h and confirm the presence of mesopores. **Figure 8** shows the presence of intra-particle pores that appear in lighter areas of the dark plane. Furthermore, the formation of inter-particle porosity was observed from the interaction of ZSM-5 crystals forming pores with a diameter of ~31.4 and 51.5 nm. The results are in accordance with the data obtained from the N_2 adsorption-desorption analysis.

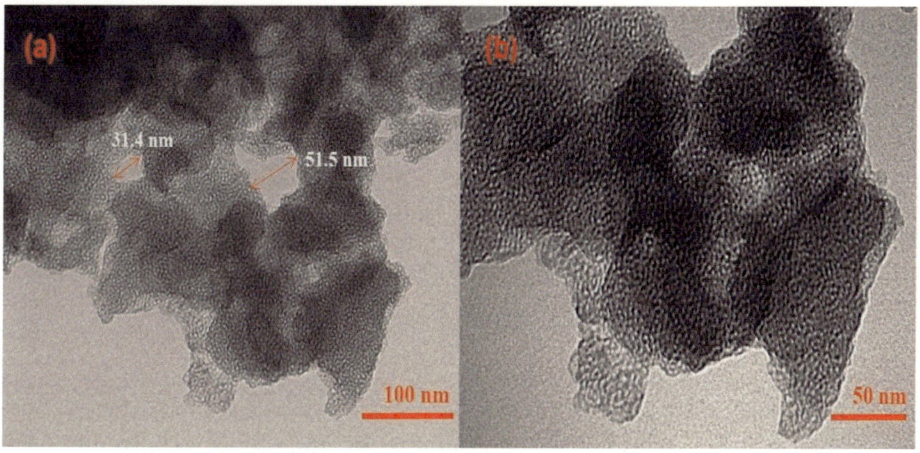

Figure 8.
TEM analysis of ZSM-5_6 h samples.

2.5 MB adsorption on ZSM-5 hierarchy

MB adsorption was carried out with RM, ZSM-5_6 h, ZSM-5_9 h, and ZSM-5_24 h to explain the effect of microporosity on MB adsorption efficiency. The adsorbents used were 2.5, 5, 10, and 15 mg, while the concentration of MB solution used was 20 mgL^{-1} shown in **Figure 9**.

A significant difference in the percentage of MB removal occurred in the number of adsorbents of 2.5 mg with ZSM-5_6 h showing *the* highest MB removal of 80%, followed by ZSM-5_9 h (68%) and ZSM-5_24 h (48%), while adsorption with RM only showed MB removal of 5%. An increase in the number of adsorbents to 5 mg led to an increase in MB removal efficiency to 95% at ZSM-5_6 h and ZSM-5_9 h, while for ZSM-5_24 h showed MB removal of 70%.

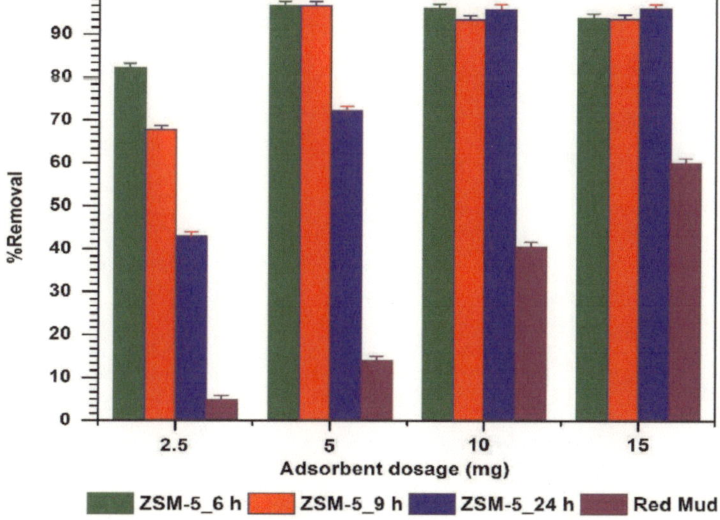

Figure 9.
Initial experiment of MB adsorption on RM and ZSM-5 hierarchy.

MB adsorption by RM increases with an increase in the amount of RM used during the adsorption process by 15 mg to obtain MB removal of up to 60%. In general, a large adsorbent surface area provides more adsorption side for MB. At a low number of adsorbents, the ZSM-5_6 h with its large surface area and mesoporous volume increases MB adsorption due to the availability of many active sides for MB adsorption. However, the binding of the number of adsorbents did not show a significant mesopore effect because the number of available adsorption sides was offset by the number of adsorbents.

2.5.1 Effect of contact time and initial concentration of MB on adsorption

The effect of the initial dye concentration on adsorption capacity was performed with a mass of 5 mg for a ZSM-5_6 h sample and the concentration of MB 20, 25, 30, and 35 mgL^{-1} solution are as illustrated in **Figure 10**, and the use of a 20 mg/L MB solution results in rapid adsorption, reaching equilibrium within 5 minutes. Higher MB concentrations lead to an increased adsorption capacity, achieving equilibrium at 30 minutes. These findings indicate that adsorption occurs swiftly at low MB concentrations due to interactions on the external surface of ZSM-5. As the MB concentration rises, MB molecules diffuse into the ZSM-5 hierarchy, eventually saturating the external surface. The repulsion between adsorbed MB molecules and those in the solution may contribute to the longer equilibrium time. However, the thrust of the high concentration of MB solution overcomes the restriction of mass transfer between the MB solution phase and the adsorbent solid. The existence of mesopore in ZSM-5_6h with a diameter of 3 nm, allows the diffusion of MB molecules into the pores, which leads to an increase in adsorption capacity at high concentrations.

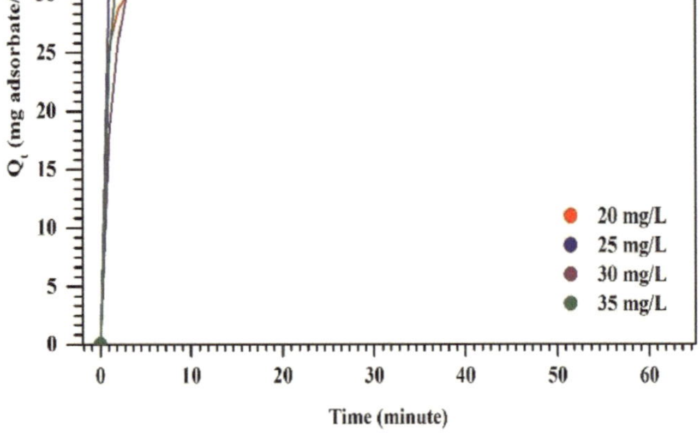

Figure 10.
Effect of initial concentration on MB adsorption using ZSM-5_6 h.

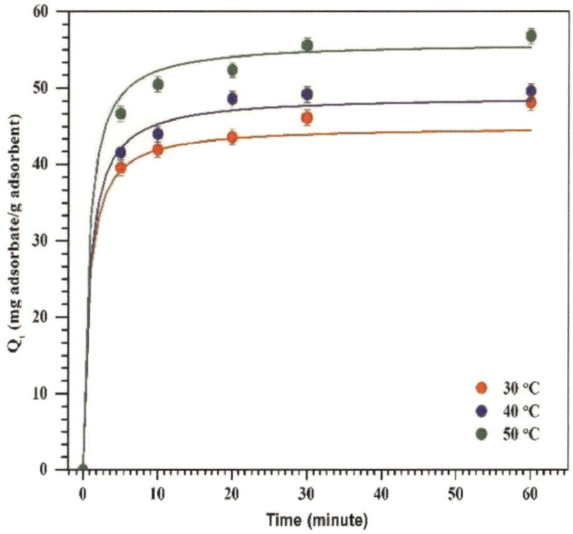

Figure 11.
Effect of temperature on MB adsorption using ZSM-5_6 h adsorbent.

2.5.2 Temperature influence

The influence of MB solution temperature during the adsorption process causes an increase in kinetics and adsorption capacity. The adsorption process of MB 20 mgL^{-1} solution at ZSM-5_6 h hierarchical is determined at temperatures of 30, 40, and 50°C. The adsorption process is also determined at 5, 10, 20, 30, and 60 min with a constant stirring speed of 2000 rpm. The adsorption capacity increases with increasing adsorption temperature as shown in **Figure 11**.

In general, MB adsorption in adsorbent solids reaches equilibrium when the adsorbent surface is covered with a single layer of MB. Elevated temperatures enhance adsorption capacity by strengthening the electrostatic interactions between MB and ZSM-5. Water molecules also engage in dipole-dipole interactions with ZSM-5, potentially occupying adsorption sites. Additionally, higher temperatures reduce adsorbate-solvent interactions, leading to stronger intermolecular forces between adsorbent particles. Conversely, at lower temperatures, electrostatic interactions between MB and ZSM-5 cause water molecules to adhere to the adsorption side. The increase in adsorption capacity at high temperatures indicates that the MB adsorption process is endothermic.

2.6 Isothermal adsorption

The adsorption of dyes in zeolite materials is an adsorption mechanism that occurs on the surface of the material. It needs to be observed experimentally to see the phenomenon of adsorption isotherms (adsorption at constant temperatures). Adsorption isothermal studies can describe how adsorbates interact with absorbents. The isotherm describes the relationship between dye concentration in the solution and the amount of dye adsorbed on the solid phase when equilibrium is achieved between the two phases.

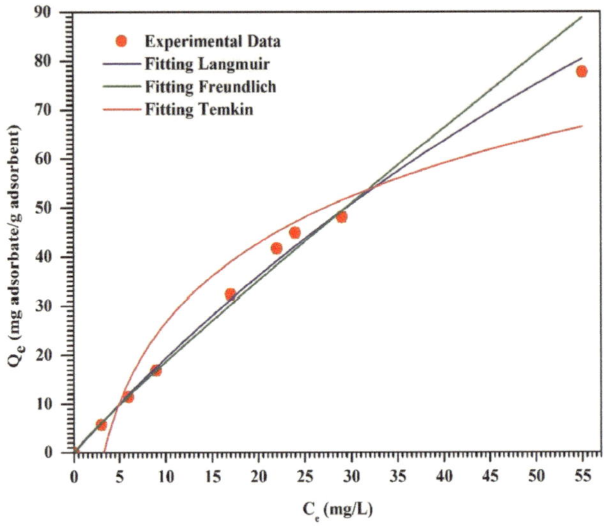

Figure 12.
Isothermal adsorption ZSM-5_6 clock using model isotherms: Langmuir, Freundlich, and Temkin.

Adsorption isotherms were evaluated using the Langmuir, Freundlich, and Temkin models to illustrate the mechanism of MB adsorption onto ZSM-5, as presented in **Figure 12**.

According to Langmuir's isotherm, adsorption involves the formation of a monolayer on the surface, assuming no interaction occurs between adsorbate molecules. The Freundlich isotherm, commonly used for heterogeneous solid catalysts, provides a mathematical model for multilayer adsorption. Meanwhile, the Temkin isotherm describes adsorption in heterogeneous systems by assuming that the heat of adsorption decreases linearly over a broad range due to interactions between adsorbent and adsorbate, with adsorption-binding energies uniformly distributed [69].

The non-linear equations of the Langmuir, Freundlich, and Temkin models [70] are shown in the Eqs. (3)–(5):

$$Q_e = \frac{Q_t K_L C_e}{1 + K_L C_e} \tag{3}$$

$$Q_e = K_F (C_e)^{1/n} \tag{4}$$

$$Q_e = B \ln K_T + B \ln C_e \tag{5}$$

where *Ce* (mg L^{-1}) represents the concentration of MB at equilibrium, while *Qe* (mg g^{-1}) denotes the amount of MB adsorbed at equilibrium. *Qe* (mg g^{-1}) is the maximum adsorption capacity based on the Langmuir model, with *KL* (L mg^{-1}) as the Langmuir constant. The Freundlich model uses *KF* (mg g^{-1}) (mg $L^{-1})^{-1/n}$ as a constant and n as a dimensionless parameter reflecting adsorption intensity. Meanwhile, *KT* (L mg^{-1}) represents the Temkin constant *KT* (L mg^{-1}), and *B* (RT/bT, $Lmol^{-1}$) is Temkin's constant with respect to adsorption heat/heat.

The MB adsorption isotherms at ZSM-5_6 h were analyzed using the Langmuir model (Eq. (3)) and the Freundlich model (Eq. (4)). Isothermal adsorption was observed at temperatures of 30, 40, and 50°C. The results of the adsorption isothermal analysis of MB at ZSM-5_6 h are shown in **Table 5**.

Isotherm model	Parameter	Value
Langmuir	Q_m (mg/g)	270.27
	K_L (L/mg)	0.008
	$R_{L,\ rata\text{-}rata}$	0.867
	RSME	0.042
Freundlich	n $(mg/g)(L/mg)^{1/n}$	1.090
	K_F $((mg\ g^{-1})(mg\ L^{-1})^{-1/n}))$	2.246
	RMSE	0.093
Temkin	B (L/mol)	23.438
	KF $((mg\ g^{-1})(mg\ L^{-1})^{-1/n}))$	0.310
	RMSE	7.597

Table 5.
Isothermal model parameters for MB adsorption at ZSM-5_6 h.

Table 5 also displays all the parameters that have been calculated from the isotherm model. The root mean square error (RMSE) value of Langmuir is 0.042, which is lower than the RMSE value of Freundlich and Temkin. These results show that MB adsorption at ZSM-5_6 h conforms to the Langmuir isotherm model with a maximum adsorption capacity of 270.27 mg g^{-1}. Based on the Langmuir model, the adsorption of MB onto the ZSM-5 surface occurs through the formation of a monolayer. The adsorption properties of MB in ZSM-5 using the Langmuir model can also be presented with a separation factor (*RL)* calculated according to Eq. (6):

$$R_L = \frac{1}{1 + K_L C_{04.4}} \tag{6}$$

The mean separation factor (*RL*) determined has a value between 0 and 1 which indicates that the adsorption properties of the Langmuir model are preferred [71].

2.7 Kinetika adsorpsi

Adsorption kinetics were determined by using Lagergren's kinetic model for pseudo-first order (**Figure 13a,c**) and pseudo-second order (**Figure 13b,d**). These two kinetic models are expressed in non-linear and linear shapes following Eqs. (7) and (8) [72] and linear following Eqs. (9) and (10) [73] as follows:

$$Q_t = Q_e\left(1 - e^{-k_1 t}\right) \tag{7}$$

$$ln\,(Q_e - Q_t) = ln\,Q_e - k_1 t \tag{8}$$

$$Q_t = \frac{Qe^2\,k_2 t}{(1 + Q_e k_2 t)} \tag{9}$$

$$\frac{t}{Q_t} = \frac{1}{k_2 Q_e^2} + \frac{t}{Q_e} \tag{10}$$

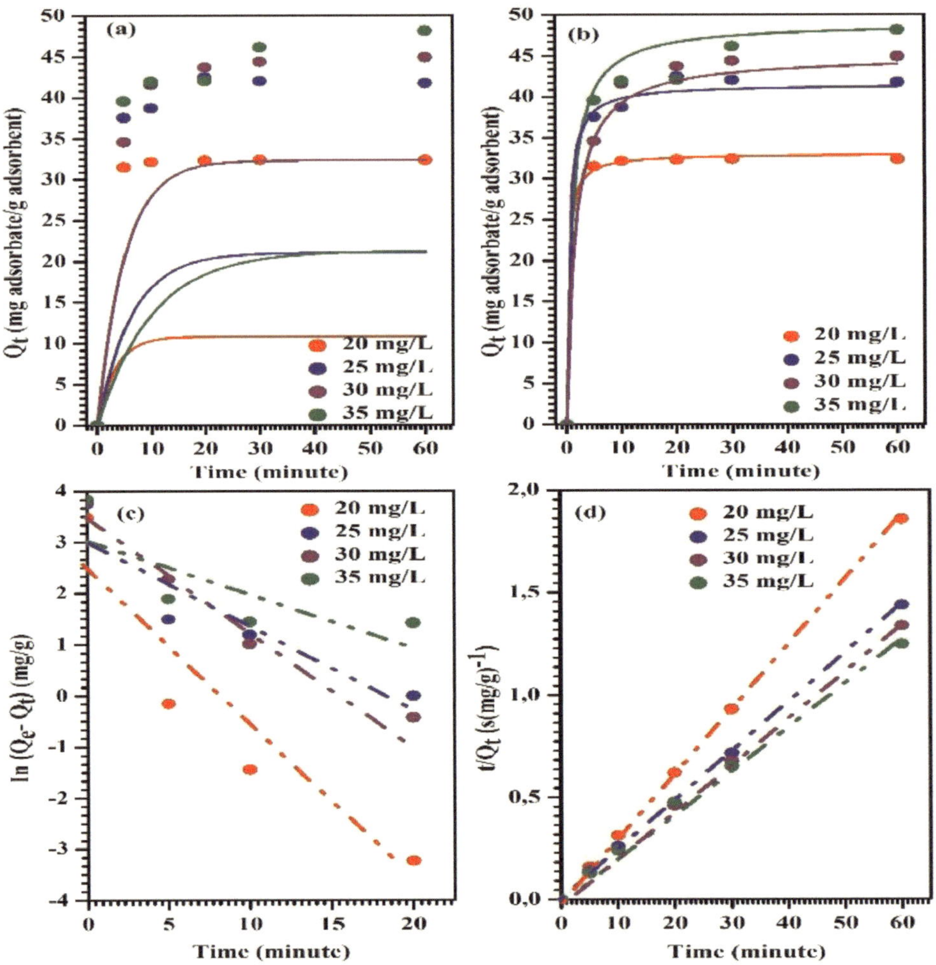

Figure 13.
Pseudo-order (a,c), MB adsorption kinetics on ZSM-5 pseudo second order (b,d), fitting non-linear (a, b), fitting linear (c, d).

where *K1* (min^{-1}) is the rate constant of the pseudo-first order and *K2* ($g\ mg^{-1}\ min^{-1}$) is the rate constant of the pseudo-second order.

The MB adsorption kinetics on the ZSM-5_6 h are shown in **Figure 13**. **Figure 13(b,d)** shows that the MB adsorption kinetics on the ZSM-5_6 h follow a - pseudo-second-order model. The Qe value calculated from the pseudo-second order corresponds to the experimental Qe compared to the Qe calculated by the pseudo-first order. All parameters of adsorption kinetics are shown in **Table 6**.

2.8 Adsorption thermodynamics

Thermodynamic parameters such as the equilibrium distribution constant (Kd) is used to determine the entropy change (ΔS°), enthalpy change (ΔH°), and Gibbs free

C_o (mg/L)	$Q_{e,\,exp}$ (mg/g)	Pseudo-order			Pseudo second order		
		k_1 (min^{-1})	Q_e (mg/g)	R^2	k_2 ($g.mg^{-1}\,min^{-1}$)	Q_e (mg/g)	R^2
20	32.32	−0.31	10.81	0.87	0.50	32.36	1.00
25	42.40	−0.16	21.07	0.84	0.07	42.01	0.99
30	44.87	−0.20	32.29	0.96	0.02	45.45	0.99
35	48.09	−0.10	21.29	0.60	0.01	48.54	0.99

Table 6.
Kinetic parameters for the pseudo-first order and pseudo-second-order models of MB adsorption on ZSM-5_6 h.

T (K)	$\Delta G°$ ($kJ\,mol^{-1}$)	$\Delta H°$ ($kJ\,mol^{-1}$)	$\Delta S°$ (kJ /mol. K)
303	−2.69	38.06	0.13
313	−4.04		
323	−5.38		

Table 7.
Thermodynamic parameters for MB dye adsorption at ZSM-5_6 h at a standard temperature of 298 K.

energy change (ΔG°). Kd is calculated using Eq. (11) based on adsorption at varying temperatures. Cads,e (mgL^{-1}) refers to the concentration of MB adsorbed at equilibrium and *Ce* (mgL^{-1}) is the concentration of MB in solution at equilibrium. The slope and intercept values of the Ln Kd graph to 1/T are used to calculate the values *of* ΔG° Eq. (12) and ΔH° Eq. (13), and ΔS° is calculated by Eq. (14).

$$K_d = \frac{C_{ad,e}}{C_e} \tag{11}$$

$$\ln K_{d0} = -\frac{\Delta G^0}{RT_0} = -\frac{\Delta H^0 - T_0 \Delta S^0}{RT_0} = \frac{\Delta S^0}{R} - \frac{\Delta H^0}{RT_0} \tag{12}$$

$$\ln\left[\frac{K_d}{K_{d0}}\right] = \frac{-\Delta H_0}{R}\left[\frac{1}{T} - \frac{1}{T_0}\right] \tag{13}$$

$$\Delta G^0 = \Delta H^0 - T_0 \Delta S^0 \tag{14}$$

The calculated thermodynamic parameters can be seen in **Table 7** showing negative values *of* $\Delta G\,°$ with increasing temperature, indicating that MB adsorption on ZSM-5 is a spontaneous process [74].

The positive ΔH° value of +38.06 kJ/mol confirms that MB adsorption onto ZSM-5 is an endothermic process, favoring higher temperatures to enhance adsorption capacity. The entropy change (ΔS°) was determined to be +0.13 kJ/mol. K shows an increase in freedom from adsorbed specimens during surface adsorption [75]. The value of entropy change was very low which indicated that there was no significant change in the resulting entropy [76, 77].

3. Conclusion

In this study, the hierarchical ZSM-5 was successfully synthesized from Red Mud (RM) waste obtained from Bintan Island, Indonesia. The synthesis process involves an alkaline fusion method to extract Al_2O_3 and SiO_2, followed by a two-stage hydrothermal crystallization process. The characterization results confirm the formation of hierarchical ZSM-5 with a combination of micropore and mesopori structures, which significantly increase its adsorption capacity. XRD analysis showed that the synthesized ZSM-5 showed well-defined crystallinity, while FTIR spectroscopy confirmed the presence of a key functional group typical of ZSM-5.

The adsorption performance of ZSM-5 hierarchically was evaluated for the removal of methylene blue dye (MB). Among the synthesized samples, ZSM-5_6 h showed the highest adsorption efficiency, eliminating up to 95% MB at a 5 mg adsorbent dose. The adsorption process follows the Langmuir isothermal model, showing the mechanism of monolayer adsorption. Kinetic studies show that adsorption follows a second-order pseudo-order model, implying chemisorption as the dominant interaction. Furthermore, thermodynamic analysis revealed that the adsorption process is spontaneous and endothermic, with increased efficiency at higher temperatures.

The findings of this study highlight the potential utilization of RM waste as a raw material for the synthesis of hierarchical ZSM-5, which can function as an efficient adsorbent for wastewater treatment. This approach not only provides an effective method to reduce RM waste but also contributes to sustainable environmental management. Future research should focus on optimizing the synthesis conditions to further increase adsorption capacity and explore the potential for regeneration of synthesized materials for long-term applications.

Author details

Hellna Tehubijuluw[1*], Didik Prasetyoko[2], Yuly Kusumawati[2], Riki Subagyo[2] and Reva Edra Nugraha[3,4]

1 Department of Chemistry, Faculty of Science and Technology, Pattimura University, Jl. Ir. M. Putuhena, Kampus Poka, Ambon, Indonesia

2 Department of Chemistry, Faculty of Sciences, Sepuluh November Institute of Technology Keputih Sukolilo, Surabaya, Indonesia

3 Department of Chemical Engineering, Faculty of Engineering, Universitas Pembangunan Nasional "Veteran" Jawa Timur, Surabaya, East Java, Indonesia

4 Low Carbon Technologies Research Center, Universitas Pembangunan Nasional "Veteran" Jawa Timur, Surabaya, East Java, Indonesia

*Address all correspondence to: hellnatehubijuluw@gmail.com

References

[1] Ajibade FO, Adelodun B, Lasisi KH, Fadare OO, Ajibade TF, Nwogwu NA, et al. Environmental Pollution and Their Socioeconomic Impacts. In Microbe Mediated Remediation of Environmental Contaminants. INC. New York, NY, United States: Woodhead Publishing (an imprint of Elsevier); 2021. DOI: 10.1016/b978-0-12-821199-1.00025-0

[2] Archambo M, Kawatra SK. RM: Fundamentals and new avenues for utilization. Mineral Processing and Extractive Metallurgy Review. 2020;**42**: 427-450. DOI: 10.1080/08827508.2020.1781109

[3] Chen X, Li X, Wu P, Zha X, Liu Y, Wei T, et al. The development and utilization of bauxite resources in the Guizhou Province and relevant challenges to the ecology and the environment. Gospodarka Surowcami Mineralnymi – Mineral Resources Management. 2022;**38**(2):5-30. DOI: 10.24425/gsm.2022.141667

[4] Gonçalves NPF, Olhero SM, Labrincha JA, Novais RM. 3d-printed red mud/metakaolin-based geopolymers as water pollutant sorbents of methylene blue. Journal of Cleaner Production. 2023;**383**:1-12

[5] Shu J, Kong D, Wang L, Khan N, Zhou S, Han Y, et al. Performance and microstructure of red-mud-blended concrete under different aqueous environments. Construction and Building Materials. 2024;**447**: 137988

[6] Niu A, Lin C. Trends in research on characterization, treatment and valorization of hazardous red mud: A systematic review. Journal of Environmental Management. 2024;**351**:119614

[7] Chen J, Wang Y, Liu Z. Red mud-based catalysts for the catalytic removal of typical air pollutants: A review. Journal of Environmental Sciences. 2023; **127**:628-640

[8] Saveliev SG, Yarosh TP, Kondratenko MM, Babaievska OV, Baboshko DY. Current state and prospects of red mud utilisation: A review. Earth and Environmental Science. 2024;**1415**:1-12

[9] Spatolisano E, Restelli F, Pellegrini LA, de Angelis AR. Waste to H2 sustainable processes: A review on H2S valorization technologies. Energies. 2024;**620**:17

[10] Rai S, Bahadure S, Chaddha MJ, Agnihotri A. Disposal practices and utilization of RM (bauxite residue): A review in Indian context and abroad. Journal of Sustainable Metallurgy. 2020; **6**(1):1-8. DOI: 10.1007/s40831-019-00247-5

[11] Nayak KC, Pathania A, Pathania AR. Red mud: Characteristics, utilization, and environmental remediation strategies in the aluminium industry. Materials Today: Proceedings. 2024;**420**:1-13

[12] Akhtar MS, Ali S, Zaman W. Innovative adsorbents for pollutant removal: Exploring the latest research and applications. Molecules. 2024;**4317**:29

[13] Kazak O, Tor A. In situ preparation of magnetic hydrochar by co-hydrothermal treatment of waste vinasse with RM and its adsorption property for Pb(II) in

aqueous solution. Journal of Hazardous Materials. 2020;**393**:122391. DOI: 10.1016/j.jhazmat.2020.122391

[14] Li B, Wu H, Liu X, Zhu T, Liu F, Zhao X. Simultaneous removal of SO_2 and NO using a novel method with RM as absorbent combined with O_3 oxidation. Journal of Hazardous Materials. 2020;**392**:122270. DOI: 10.1016/j.jhazmat.2020.122270

[15] Qi X, Wang H, Zhang L, Xu B, Shi Q, Li F. Removal of Cr (III) from aqueous solution by using BauxiteResidue (RM): Identification of active components and column tests. Chemosphere. 2020;**245**: 125560. DOI: 10.1016/j.chemosphere.2019.125560

[16] Shi W, Ren H, Huang X, Li M, Tang Y, Guo F. Low cost RM modified graphitic carbon nitride for the removal of organic pollutants in wastewater by the synergistic effect of adsorption and photocatalysis. Separation and Purification Technology. 2020;**237**: 116477. DOI: 10.1016/j.seppur.2019.116477

[17] Padhi MK, Nanda BB, Kar P, Nanda B. A review on sustainable utilization of red mud based photocatalytic material for environmental pollution abetment. Materials Today: Proceedings. 2023: 2214-7853

[18] Bahari MB, Mamat CR, Jalil AA, Hassan NS, Nabgan W, Setiabudi HD, et al. Mesoporous alumina: A comprehensive review on synthesis strategies, structure, and applications as support for enhanced H2 generation via CO_2-CH_4 reforming. International Journal of Hydrogen Energy. 2022;**47**: 41507-41526

[19] Zhao C, Hu X, Liu C, Chen D, Yun J, Jiang X, et al. Hierarchical architectures of ZSM-5 with controllable mesoporous and their particular adsorption/desorption performance for VOCs. Journal of Environmental Chemical Engineering. 2022;**10**:106868

[20] Zinoveev D, Pasechnik L, Grudinsky P, Yurtaeva A, Dyubanov V. Kinetics and mechanism of red mud carbothermic reduction and reduced iron grain growth: An influence of sodium sulfate. Crystals. 2023;**13**:1

[21] Aslam MMA, Sun T, Dai M, Xu W, Ye Y, Ali I, et al. Technologies for recovery of iron from red mud: Processes, challenges and opportunities. Sustainable Materials and Technologies. 2024;**41**:e01053

[22] Kong H, Zhou T, Yang X, Gong Y, Zhang M, Yang H. Iron recovery Technology of red mud—A Review. Energies. 2022;**3830**:15

[23] Alsagheer N. Synthesis of sodium silicate from the acid leached calcined kaolinitic clay residue of aluminum sulfate industry. Research Square. 2024:1-24

[24] Lin S, Jiang X, Zhao Y, Yan J. Zeolite greenly synthesized from Fly ash and its resource utilization: A review. Science of the Total Environment. 2022;**851**:158182

[25] Zhang Z, Wang Y, Zhang Y, Shen B, Ma J, Liu L. Stabilization of heavy metals in municipal solid waste incineration fly ash via hydrothermal treatment with coal fly ash. Waste Management. 2022;**144**: 285-293

[26] Sun Z, Shu Q, Zhang Q, Li S, Zhu G, Wang C, et al. A hydrothermal synthesis process of ZSM-5 zeolite for VOCs adsorption using desilication solution. Separations. 2024;**39**:11

[27] Aloui L, Mezghich S, Mansour L, Hraiech S, Ayari F. Swift removal of the heavy metals cadmium and lead from an

aqueous solution by a CANZeolite synthesized from natural clay. ChemEngineering. 2023;**113**:7

[28] Aguirre-Cruz G, Legorreta-Garcia F, Aguirre-Cruz G, Stanciu L, Aguirre-Alvarez G. Synthesis of hierarchical silica zeolites for heterogenous catalysis and adsorption. Microporous and Mesoporous Materials. 2022;**345**:112274

[29] Hamid A, Reva EN, Holilah H, Hasliza B, Prasetyoko D. Large intraparticle mesoporosity of hierarchical ZSM-5 synthesized from kaolin using silicalite seed: Effect of aging time and temperature. Journal of the Korean Ceramic Society. 2022:1-13

[30] Ren X-Y, Cao J-P, Ma M-W, Zhao X-Y. Insight on product regulation and catalysts deactivation during catalytic fast pyrolysis of lignite over micro/mesoporous catalysts synthesized by ZSM-5 seed. Journal of Analytical and Applied Pyrolysis. 2024; **179**:1-26

[31] da Rocha JDG, Macuvele DL, Aneo P, de Andrade CJ, Riella HG, Padoin N, et al. Advances and environmental aspects on the synthesis of hierarchical zeolites revisited: A state-of-the-art description. Journal of Environmental Chemical Engineering. 2023;**11**(2):109397

[32] Abdulkadir BA, Zaki RSRM, Wahab ATA, Miskan SN, Nguyen A-T, Vo D-VN, et al. A concise review on surface and structural modification of porous zeolite scaffold for enhanced hydrogen storage. Chinese Journal of Chemical Engineering. 2024;**70**:33-53. DOI: 10.1016/j.cjche.2024.03.001

[33] Martins A, Nunes N, Carvalho AP, Martins LMDRS. Zeolites and related materials as catalyst supports for hydrocarbon oxidation reactions. Catalysts. 2022;**154**:1-52

[34] Ahmed MJ, Hameed BH, Hummadi EH. Review on recent progress in chitosan/chitin-carbonaceous material composites for the adsorption of water pollutants. Carbohydrate Polymers. 2020;**247**:116690. DOI: 10.1016/j.carbpol.2020.116690

[35] Akhtar MS, Ali S, Zaman W. Innovative adsorbents for pollutant removal: Exploring the latest research and applications. Molecules. 2024; **4317**:29

[36] Gkika DA, Mitropoulos AC, Kyzas GZ. Why reuse spent adsorbents? The latest challenges and limitations. Science of The Total Environment. 2022; **82**:153612. DOI: 10.1016/j.scitotenv.2022.153612

[37] Li G, Liu J, Yi L, Luo J, Jiang T. Bauxite residue (Redmud) treatment: Current situation and promising solution. Science of the Total Environment. 2024;**948**:174757

[38] Powolny T, Dumańska-Słowik M, Szczerbowska-Boruchowska M, Woszczyna MTT. Water and Moganite Distribution in Agates From Bou Hamza, Morocco: Constraints From A Combined Microscopic And Spectroscopic Study. 1-27

[39] Ghosh S, Mahato S, Dutta T, Ahamed Z, Ghosh P, Roy P. Highly selective, sensitive and biocompatible rhodamine-based isomers for Al^{3+} detection: A comparative study. Spectrochimica Acta Part A: Molecular and Biomolecular Spectroscopy. 2024; **305**:123455

[40] Liu X, Zhang N, Sun H, Zhang J, Li L. Structural investigation relating to the cementitious activity of bauxite

residue - RM. Cement and Concrete Research. 2011;**41**(8):847-853. DOI: 10.1016/j.cemconres.2011.04.004

[41] Gök A, Omastová M, Prokeš J. Synthesis and characterization of RM/polyaniline composites: Electrical properties and thermal stability. European Polymer Journal. 2007;**43**(6): 2471-2480. DOI: 10.1016/j.eurpolymj.2007.03.005

[42] Carneiro J, Tobaldi DM, Hajjaji W, Capela MN, Novais RM, Seabra MP, et al. RM as a substitute coloring agent for the hematite pigment. Ceramics International. 2018;**44**(4):4211-4219. DOI: 10.1016/j.ceramint.2017.11.225

[43] Ramdhani EP, Santoso E, Holilah H, Reva EN, Hasliza B, Suprapto S, et al. Direct synthesis of Fe-aluminosilicates from red mud for catalytic deoxygenation of waste cooking oil. RSC Advances. 2023;**13**:31989. DOI: 10.1039/d3ra05910c

[44] Ndlovu NZ, Ameh AE, Petrik LF, Ojumu TV. Synthesis and characterisation of pure phase ZSM-5 and sodalite zeolites from coal fly ash. Materials Today Communications. 2023; **34**:105436. DOI: 10.1016/j.mtcomm.2023.105436

[45] Ou Y, Ma S, Wang X, Wu H, Hui H, Zhang Y. Dissolution behavior of aluminum and silicon from kaolinite diaspore symbiotic low-grade bauxite in low alkali system. Colloids and Surfaces A: Physicochemical and Engineering Aspects. 2024;**701**:134821. DOI: 10.1016/j.colsurfa.2024.134821

[46] Zhou G, Wang Y, Qi T, Zhou Q, Liu G, Peng Z, et al. Toward sustainable green alumina production: A critical review on process discharge reduction from gibbsitic bauxite and large-scale applications of red mud. Journal of Environmental Chemical Engineering. 2023;**11**(2):109433. DOI: 10.1016/j.jece.2023.109433

[47] Li Q, Dong M, Li R, Cui Y, Xie G, Wang X, et al. Enhancement of Cr(VI) removal efficiency via adsorption/photocatalysis synergy using electrospun chitosan/g-C_3N_4/TiO_2 nanofibers. Carbohydrate Polymers. 2021;**253**:117200. DOI: 10.1016/j.carbpol.2020.117200

[48] Kusumawardani LJ, Iryani A, Sinaga, and Ennita L. Modification of zeolite made from coal fly ash with TiO_2: Effect of aging time on physical and optical properties. Makara Journal of Science. 2023;**27**(1). Article 8. DOI: 10.7454/mss.v27i1.1483. Available from: https://scholarhub.ui.ac.id/science/vol27/iss1/8

[49] Liu Q, Li X, Cui M, Wang J, Lyu X. Preparation of eco-friendly one-part geopolymers from gold mine tailings by alkaline hydrothermal activation. Journal of Cleaner Production. 2021;**298**:126806. DOI: 10.1016/j.jclepro.2021.126806

[50] Nguyen K, Dinh P, Nguyen HQ, Hung NT. Zeolite ZSM-5 synthesized from natural silica sources and its applications: A critical review. Journal of Chemical Technology & Biotechnology. 2023;**98**(6):1339-1355. DOI: 10.1002/jctb.7380

[51] Gao Y, Zhao W, Zhang D. Preparation of prismatic g-C_3N_4 by CTAB hydrothermal method and its degradation performance under visible light. Diamond and Related Materials. 2021;**121**:108787. DOI: 10.1016/j.diamond.2021.108787

[52] Yinchang P, Yijian Z, Qinglin X, Nanchun C. Two-step hydrothermal synthesis and conversion mechanism of zeolite X from stellerite zeolite. RSC Advances. 2022;**12**:3313. DOI: 10.1039/d1ra07798h

[53] Abubakar A, Nasiru YP, Wilson LD. Synthesis and characterization of zeolite A from Alkaleri Kaolin using conventional hydrothermal synthesis technique. Nanochem Res. 2023;**8**(3):190-196

[54] Ou Z, Li Y, Wu W, Bi Y, Xing E, Yu T, et al. Encapsulating subnanometric metal clusters in zeolites for catalysis and their challenges. Chemical Engineering Journal. 2022;**430**:132925. DOI: 10.1016/j.cej.2021.132925

[55] Feng Z, Yuan R, Wang F, Chen Z, Zhou B, Chen H. Preparation of magnetic biochar and its application in catalytic degradation of organic pollutants: A review. Science of the Total Environment. 2021;**765**:142673. DOI: 10.1016/j.scitotenv.2020.142673

[56] Shakeri M, Khatami Shal Z, Van Der Voort P. An overview of the challenges and progress of synthesis, characterization and applications of plugged SBA-15. Materials for Heterogeneous Catalysis. Materials. 2021;**14**:5082. DOI: 10.3390/ma14175082

[57] Melek KK, Ikbal SY, Selçuk B, Münevver S. Role of a novel cationic gemini surfactant (CGS) on a one-step sol–gel process and photocatalytic properties of TiO_2 powders. Journal of Water and Health. 2022;**20**(11):1629. DOI: 10.2166/wh.2022.069

[58] Yang S, Lei G, Xu H, Lan Z, Wang Z, Gu H. Metal oxide based heterojunctions for gas sensors: A review. Nanomaterials. 2021;**11**(4):1026. DOI: 10.3390/nano11041026

[59] Jia H, Tao D, Xin F, He G, Ziyang Q, Yingnan L, et al. Synthesis of template-free ZSM-5 from Rice husk ash at low temperatures and its CO_2 adsorption performance. ACS Omega. 2022;**6**:3961-3972. DOI: 10.1021/acsomega.0c05842

[60] Lin F, Ye Z, Kong L, Liu P, Zhang Y, Zhang H, et al. Facile morphology and porosity regulation of zeolite ZSM-5 Mesocrystals with synergistically enhanced catalytic activity and shape selectivity. Nanomaterials. 2021;**12**(9):1601. DOI: 10.3390/nano12091601

[61] Nakhaei Pour A, Mohammadi A. Kinetic study of the crystallization of ZSM-5 under organic template-free conditions. Inorganic and Nano-Metal Chemistry. 2021;**53**(1):33-38. DOI: 10.1080/24701556.2021.1978494

[62] Zhou L, Wang M, Yang S, Guo W, Pu X, He Y, et al. Facile synthesis of mesoporous ZSM-5 aided by sonication and its application for VOCs capture. Ultrasonics Sonochemistry. 2022;**88**:106098. DOI: 10.1016/j.ultsonch.2022.106098

[63] Feng M, Kou Z, Tang C, Shi Z, Tong Y, Zhang K. Recent progress in synthesis of zeolite from natural clay. Applied Clay Science. 2023;**243**:107087. DOI: 10.1016/j.clay.2023.107087

[64] Shu Q, Sun Z, Zhu G, Wang C, Li H, Qi F, et al. Highly efficient synthesis of ZSM-5 zeolite by one-step microwave using desilication solution of coal gasification coarse slag and its application to VOCs adsorption. Process Safety and Environmental Protection. 2022;**167**:173-183. DOI: 10.1016/j.psep.2022.08.068

[65] Fatma KD. Investigation of the effect of different synthesis methods on the photocatalytic activity of TiO_2: Comparison of rutile and Anatase TiO_2. Cumhuriyet Science Journal. 2022;**43**(3):409-415. DOI: 10.17776/csj.1107688

[66] Velaga B, Doley R, Peela NR. Rapid synthesis of hierarchical ZSM-5 zeolites for the reactions involving larger reactant molecules. Advanced Powder

Technology. 2021;**32**(4):1033-1046. DOI: 10.1016/j.apt.2021.02.002

[67] Feng M, Kou Z, Tang C, Shi Z, Tong Y, Zhang K. Recent progress in synthesis of zeolite from natural clay. Applied Clay Science. 2023;**243**: 107087. DOI: 10.1016/j.clay.2023.107087

[68] Xu X, Bao S, Zhang Y, Ping Y. Sustainable enhancement of fly ash-based geopolymers: Impact of alkali thermal activation and particle size on green production. Process Safety and Environmental Protection. 2024;**191**: 478-489. DOI: 10.1016/j.psep.2024.08.133

[69] Ighnih H, Haounati R, Ouachtak H, Regti A, El Ibrahimi B, Hafid N, et al. Efficient removal of hazardous dye from aqueous solutions using magnetic kaolinite nanocomposite: Experimental and Monte Carlo simulation studies. Inorganic Chemistry Communications. 2023;**153**:110886. DOI: 10.1016/j.inoche.2023.110886

[70] Ma X, Duan D, Chen J, Xie B. Structure and adsorption performance of cationic *Entermorpha prolifera* polysaccharide-based hydrogel for typical pollutants: Methylene blue, cefuroxime, and Cr (VI). Gels. 2022; **8**(9):546. DOI: 10.3390/gels8090546

[71] Saxena M, Sharma N, Saxena R. Highly efficient and rapid removal of a toxic dye: Adsorption kinetics, isotherm, and mechanism studies on functionalized multiwalled carbon nanotubes. Surfaces and Interfaces. 2020;**21**:100639. DOI: 10.1016/j.surfin.2020.100639

[72] Chinni SV, Sanniraj A, Koduru JR, Lebaka VR. Nanobioremediation—An emerging eco-friendly approach for a sustainable environment. Hybrid Nanomaterials for Sustainable Applications. 2023:199-219. DOI: 10.1016/B978-0-323-98371-6.00012-4

[73] Faisol M, Muhammad BS, Mohamed HG, Mamdouh TJ, Norma AF, Moch S. Lead (Pb) biosorption by intact biomass and alginate extract of *Sargassum crassifolium* originated from Gresik regency waters Indonesian. Journal of Marine Sciences. 2022;**27**(2): 159-168

[74] Mulushewa Z, Dinbore WT, Ayele Y. Removal of methylene blue from textile waste water using kaolin and zeolite-x synthesized from Ethiopian kaolin. Environmental Analysis, Health and Toxicology. 2021;**36**(1):e2021007. DOI: 10.5620/eaht.2021007

[75] You J, Zhang L, He L, Zhang B. Photocatalytic degradation of methyl orange on $ZnO–TiO_2/SO_4^{2-}$ heterojunction composites. Optical Materials. 2022;**131**:112737. DOI: 10.1016/j.optmat.2022.112737

[76] Li D, Sun L, Yang L, Liu J, Shi L, Zhuo L, et al. Adsorption behavior and mechanism of modified *Pinus massoniana* pollen microcarriers for extremely efficient and rapid adsorption of cationic methylene blue dye. Journal of Hazardous Materials. 2024;**465**: 133308. DOI: 10.1016/j.jhazmat.2023.133308

[77] Ying Z, Huang L, Ji L, Li H, Liu X, Zhang C, et al. Efficient removal of methylene blue from aqueous solutions using a high specific surface area porous carbon derived from soybean dreg. Materials. 2020;**14**(7):1754. DOI: 10.3390/ma14071754